The New Radio Receiver Building Handbook

And Related Radio Subjects

Vacuum Tube and Transistor

Shortwave Radio Receivers

by Lyle Russell Williams, BSEE

KC5KBG

Copyright 2006 by Lyle Russell Williams
All rights reserved
lyle009@juno.com

Published in the United States by:
The Alternative Electronics Press

ISBN 978-1-84728-526-3

Copies available from lulu.com and other online book sellers.

Contents

Introduction: Who Builds Radios Today?

Chapter 1: Types of Receivers

    Discussion of the Drawings of this Book
    The Crystal Radio
    The Regenerative Radio
    The TRF Receiver
    The Superheterodyne Receiver
    The Super-regenerative Receiver
    The Direct Conversion Receiver
    Conclusion

Chapter 2: Types of Digital Receivers

    Digitally Synthesized Analog Superhet
    All Digital Receivers
    Digital Modulation
    Spread Spectrum

Chapter 3: Radio Services Currently Available

    Long Wave
    Medium Wave
    Shortwave
    FM Broadcast Band
    Above 108 mHz
    Internet Radio
    Big Dish Satellite Radio
    XM and Sirius Satellite Radio
    Summary

Chapter 4: Old Regenerative Radio Circuits

        The Grid Leak Detector
        The Original Regenerative Radio
        Frequency Dials
        Spurious Radiation
        DeForest Versus Armstrong
        Passive and Active Regeneration Control
        Problems Associated with the Regeneration Control
        Regeneration Control Methods

Chapter 5: The New Regenerative Radio

        Replacing the Tube in an Old Circuit With a Transistor
        Regenerative Radio Configurations
        Block Diagram for All Regenerative Radios
        The Regenerative Amplifier
        The Interim Amplifier
        The AM Detector
        The Audio Power Amplifier

Chapter 6: Other Circuits Using the Regenerative Principle

        RF Oscillators
        Rudimentary AM Transmitter
        The Q-multiplier
        The Regenerative IF Amplifier

Chapter 7: Mechanical Aspects of Radio Design

        Breadboard Type of Radio Construction
        Chassis Type of Construction
        Printed Circuits
        Prototyping Tube Circuits
        Semiconductor Prototyping

    Variable Capacitors
    Mechanical Reduction Drives
    Variometers and Other Inductors
    Panels
    Frequency Dials
    Knobs and Pointers
    Dial Lighting

Chapter 8: Vacuum Tube Radio Design and Components

    Availability of Parts
    Vacuum Tubes
    Early Dry Battery Tubes
    Indirectly Heated Cathode Tubes
    Rectifier Tubes and Diodes
    Transformers for Tube Designs
    AC-DC Series String Radios

Chapter 9: Semiconductor Radio Design and Components

    Bipolar and Field Effect Transistors
    Integrated Circuits
    Passive Components
    Frequency Linearity of Tuning Circuits
    Varactor Tuning Circuits

Chapter 10: Electronic Test Equipment

    Inexpensive Testing
    Personal Computers and Software
    Pricing
    Inexpensive Instruments
    Calibration Standards

Chapter 11: Complete Receivers

    The Ocean Hopper
    The Space Spanner
    The Globe-Span Modern Transistor/IC Regenerative Radio
    Further Modifications and Improvements of the Transistor/IC Radio
    Newly Designed Vacuum Tube Regenerative Radio

Tables, Drawings, and Photographs

Introduction:

Chapter 1: Types of Receivers

    TABLES:

        1-1   Frequency Ranges of the Shortwave Bands
        1-2   Toroidal Inductor Winding Instructions for the Shortwave Broadcast Bands

    FIGURES:

        1-1A   A Crystal Radio Schematic
        1-1B   A Home Constructed Crystal Radio
        1-1C   A Manufactured Crystal Radio
        1-2   A Crystal Detector in a Holder
        1-3A   Voltage Versus Current (V-I) Characteristic of a Diode
        1-3B   How the Diode Characteristic Removes the Negative Half of a Signal
        1-4   The Classical Illustration of the Detection of an AM Signal
        1-5   A Grid Leak Detector
        1-6   Regenerative Radio Similar to Armstrong's Original
        1-7   A Tuned Radio Frequency (TRF) Receiver
        1-8A   Block Diagram of a Superheterodyne Radio
        1-8B   Schematic Diagram of a Superheterodyne Receiver
        1-9A   A Super-regenerative Receiver
        1-9B   Output Waveforms of a Super-regenerative Receiver
        1-10   Block Diagram of a Direct Conversion Receiver

Chapter 2: Digital Radio

        2-1   A Digitally Synthesized Analog Receiver

Chapter 3:   Radio Media

Chapter 4: Old Regenerative Designs

    4-1   A Grid Leak Detector
    4-2A   The AM Input Signal to the Grid Leak Detector
    4-2B   The Signal at the Grid of the Grid Leak Detector
    4-3   A Vacuum Diode AM Detector
    4-4   The First Regenerative Detector
    4-5   A "Log" Dial Used for Tuning
    4-6   DeForest's Regenerative Radio
    4-7   Regeneration Control by Throttling Capacitor
    4-8   Regeneration Control by Variable Resistor
    4-9   Regeneration Control by Changing Plate Voltage
    4-10   Regeneration Control by Changing the Screen Voltage
    4-11   Regeneration Control by Changing the Control Grid Bias

Chapter 5: New Regenerative Radio Designs

    5-1   Replacing the Vacuum Tube in an Antique Circuit with a JFET
    5-2   An Alternate Method of Replacing a Tube with a JFET
    5-3   Tickler Coil Regenerative Circuit
    5-4   Tuned Collector (Tuned Plate) Regenerative Circuit
    5-5   Pierce Regenerative Circuit
    5-6   Hartley Regenerative Circuit
    5-7   Colpitts Regenerative Circuit
    5-8A   Block Diagram of a Regenerative Receiver
    5-8B   Alternate Block Diagram for Tube Regenerative Receivers
    5-9   A Colpitts Regenerative Amplifier
    5-10   A Dual Gate MOSFET Colpitts Regenerative Amplifier
    5-11   JFET Version of Colpitts Regenerative Amplifier
    5-12   A Hartley Regenerative Amplifier
    5-13   A Hartley Regenerative Amplifier with Buffer
    5-14   An Interim Wide Band RF Amplifier

5-15	A High Gain Interim Audio Amplifier
5-16	Dual Triode Interim Audio Amplifier
5-17	Detector Used in Most Commercial AM Receivers
5-18A	AM Detector For Use When Source is at Zero DC Offset Voltage
5-18B	AM Detector That Works When There is a DC Component in the Input RF Signal
5-19	Non Loading Detector That Works With an Input DC Offset
5-20	An Integrated Circuit Audio Power Amplifier
5-21	A Vacuum Tube Power Amplifier

Chapter 6: Other Circuits Using the Regenerative Principle

6-1	A Tickler Coil RF Oscillator
6-2	A Low Power AM Transmitter
6-3	A Q-Multiplier Add On Circuit
6-4	A Regenerative IF Stage in a Superheterodyne Receiver

Chapter 7: Mechanical Aspects of Radio Design

7-1	An Example of Metal Chassis Construction
7-2	A Nibbling Tool and Several Chassis Punches
7-3	A Vacuum Tube Radio on a Printed Circuit Board
7-4A	A Transistor and Integrated Circuit Printed Circuit Board
7-4B	Underside of the Printed Circuit Board Before the Parts Are Inserted
7-5	A Transistor and IC Circuit Fabricated on Vector Board
7-6	Circuit Connections on a Spring Terminal Board
7-7	Variable Tuning Capacitors
7-8	A Variometer Inductor
7-9	Various Inductors
7-10	Tuning Dial For a Nine Band Variable Capacitor Tuned Receiver
7-11	Tuning Dial For a Six Band Varactor Tuned Receiver
7-12A	Control Knobs

7-12B   Underside of Above Control Knobs

Chapter 8: Vacuum Tube Radio Design and Components

    8-1   Octal Base and Miniature Glass Vacuum Tubes

Chapter 9: Semiconductor Design

9-1	Dial From a 1936 Philco Radio
Table 9-1	Changes In Tuning Capacitance Needed to Produce Equal Changes In Frequency
9-2(A-D)	Varactor Tuning Curves
9-3	Full Details of the Simplified Circuit of Figure 9-2D

Chapter 10: Electronic Test Equipment

10-1	An Analog Volt-Ohm-Milliampere (VOM) Meter
10-2	A Digital Multi-Meter
10-3	A Circuit For Measuring an RF AC Signal on a DC Meter
10-4	A Digital Inductance-Capacitance-Resistance (LCR) Meter
10-5	An Audio Signal Generator
10-6	A Radio Frequency (RF) Generator
10-7	An Inexpensive Frequency Counter
10-8	A Homebuilt Crystal Controlled Marker Generator
10-9	A Noise Bridge For Measuring Impedance
10-10	An Oscilloscope

Chapter 11: Complete Receivers

11-1	The Ocean Hopper Radio Schematic
11-2	Three Views of the Ocean Hopper Radio
11-3	The Space Spanner Radio Schematic
11-4	The Space Spanner Radio
11-5	Schematic Diagram of Globe Span Receiver
- - -	Parts list of the Globe Span Receiver

11-6A   The Globe Span Radio Front View
11-6B   The Globe-Span Radio Rear View

11-7   Full Scale Printed Circuit Foil Side Pattern for the Globe-Span Radio
11-8   Parts Placement Diagram for Printed Circuit in Figure 11-7
11-9   Front Panel of the Globe-Span Radio
11-10   Schematic of the Tube Regenerative Radio
11-11   The Vacuum Tube Regenerative Radio with Plug-in Coils

Who Builds Radios Today?

Morse code radio transmission and reception started in the 1890s. The initial passive detectors used for code were not integrating and could not detect voice transmissions. When galena and other minerals began to be used as detectors, it became possible to demodulate voice AM signals.

The first electronic device used in radios was the vacuum tube (then called the audion) which was patented in 1908. The inventor, DeForest, substituted this amplifying device in place of the crystal detector in the radio receiver of the day. The amplifying ability of the tube was soon found to be useful in the telephone industry and in many other devices from hearing aids to industrial controls.

The transistor, which was invented in 1947, and later the integrated circuit, replaced the amplifying function of the vacuum tube in most applications. From the concept of the amplifying vacuum tube came all of today's electronic devices, including the computer. Electronic devices are a part of an amazing array of modern products such as wrist watches, automobiles, trains, airplanes, telephones, robots, and most anything else involving communications, timing, control, or computation.

The initial builders of broadcast receivers were amateurs and hobbyists. Pioneering magazines such as those published by Hugo Gernsbach, and amateur radio publications such as QST, provided design information needed to build radio receivers. Mail order companies and, in some cities, radio parts stores provided a source of radio parts for these builders.

Commercially manufactured receivers became available in the form of the three dial TRFs, and later in the cathedral radios of the 1920s. Hobbyists continued to build receivers during this decade.

From the 1930s forward, commercial interests dominated radio receiver production. In this early period, there were several hundred (some sources say more than one thousand) radio receiver manufacturers in the world.

Starting in the late 1940s, electronic kits became popular. The devices available in kit form were radio receivers, radio transmitters, test equipment, stereo/audio equipment, and even televisions. These kits included the chassis, printed circuit board, all parts, and an instruction book for building an electronic device. The kits required less expertise than building a device from a magazine article or self-designing an electronic device. The kits could be built for less money than manufactured items because the builder did his own manufacturing. This saved money because manufacturing costs at that time were high. Kit suppliers, nevertheless, incurred the costs of writing and publishing instruction manuals and of support for those who had difficulty building the circuit.

Three trends have resulted in the loss of the financial incentive for building kits. The complexity of ordinary devices such as portable phones has become enormous. The actual function of such a device may be implemented in software. No one would want to build such a device from individual parts. Parts have become smaller and many are of the surface mounting (SMT) type that are difficult to attach to the circuit board by manual processes. Many products are designed with the entire circuit on a single integrated circuit chip or on two or three chips called a chipset.

The relentless quest for productivity in manufacturing and design has greatly reduced the cost of electronic manufactured products. Moreover, most products today are manufactured by low paid workers in foreign countries. Manufacturing costs have become so low that no saving can be gained by kit builders. The famous Heathkit company that supplied electronic kits for 40 years went out of business in the 1980s.

Although the financial incentive of kit building is gone, kits are still sold as educational projects. Kits are also available for electronic devices that are not available in manufactured form. One example is the radio kit offered by the author later in this book.

Electronic devices today are remarkably inexpensive. A manufactured console radio in the 1940s cost an average person's yearly salary. Today, such a complex device as a 20-inch color television can be purchased for the equivalent of about $10 in 1940s dollars.

In the 1960s, the Japanese began to dominate the portable transistor radio market. The Japanese transistor radios were very cheap, but early designs were crude by U.S. standards. The Japanese were reluctant to make the switch from germanium to silicon transistors and they lagged behind the U.S. for a number of years. The Japanese eventually improved their designs and came to dominate the entire consumer electronics market.

The digitalization of electronics is an unpredicted phenomena. Still photos, audio recordings, and movies are recorded by converting the analog information into a series of binary numbers and then storing the numbers on CD ROMs, DVD ROMs or other digital media, or by transmitting the numbers over the air. This has become possible due to the incredible increase in the power of computers over the past few years. Nearly every electronic device today contains at least some digital circuitry.

Some hobbyists today are involved with digital projects and work with microprocessors such as the PIC and "Basic Stamp" to control anything from mousetraps to robots.

It is possible to assemble a computer by obtaining separate motherboards, video cards, audio cards, disk drives, modems, power supplies, and metal cabinets. These can be combined into a complete computer. Usually, the only tool required for this is a screwdriver. The various cards are plugged into the motherboard and the assembly can be made without soldering. The computer can be tailored to individual specifications and sometimes, money can be saved over a commercially assembled computer. There are magazine articles published that can assist in home computer assembly.

Today's technology is supplying new analog components. Traditional audio devices such as operational amplifiers have increased in frequency of operation until many can now be used at radio frequencies. Digital components called digital signal processors (DSP), can perform analog functions with precision. The DSP is a contained digital computer configured to accomplish a specific single task and it can be packaged as a separate module or plug-in card. One function that a DSP could accomplish is an IF filter with steep skirts for use in superheterodyne receivers. Analog IF filters are quite expensive and the price rises precipitously as the number of filter poles increases. If a DSP IF filter was made available to hobbyists at a reasonable price, there would be a market for the part.

It might be thought that the regenerative radio, which has been used since 1914, would have been improved as much as it could be by the present time. However, the regenerative radio has not been the subject of constant research as the superheterodyne has been. It will be seen that significant improvements of the older regenerative circuits are possible.

The concept of productivity has permeated the design world as well as the world of manufacturing. Today, engineers work on deadlines that leave little time for innovation. Designs contain digital circuitry and the actual design is done with heavy reliance on computer aided design (CAD).

Electronic design as a profession has always been problematic. Designers paid for 40 hour weeks have traditionally been expected to work many extra hours with no payment for the extra time. This is an example of "fake productivity": productivity realized by workers being forced to provide free work time.

Engineers are subject to layoffs when the economic cycle is in a decline. For many years, foreign engineers have been brought to the U.S. on work visas. The foreign engineers work for somewhat lower salaries than U.S. born engineers. This importing of engineers has been promoted by lobbyists working for the electronics companies that desire cheap design workers. Importing of foreign engineers occurs even when there is considerable unemployment in the engineering field.

A more recent trend is that engineering designs are themselves exported (out-sourced) to foreign countries, principally China and India. Engineers in China work for salaries of about one fifth of what U.S. designers make. Such salaries would be below poverty level in the U.S. It appears that while consumer electronics design has long since moved to Japan, there is a current trend toward all commercial electronic designs moving to countries in Asia.

If this trend continues, we will depend upon other countries, even for electronic devices for our military equipment. Electronics industry spokesmen argue that out-sourcing of technology is good for the U.S. It is hard to imagine how this can be true. As a nation, we once had intense pride in our technological prowess, our ability to send men and equipment into space and to the moon. U.S. engineers during this period were highly regarded.

So what is the answer to the question "Who builds radios in the U.S. today?". The answer is: America's amateurs, hobbyists and midnight engineers. In other words, us.

1 Types of Radio Receivers

Discussion of the Drawings of this Book

The schematic drawings of this book are either traditional circuits from old sources, circuits taken from amateur radio or engineering sources, or circuits developed and tested by the author. Approximate component values and vacuum tube or transistor types have been included in these drawings when possible.

While all of the circuits have not been tested by the author, it is believed that any of the circuits will work as shown or with very minor modifications. Tube circuits are more prone to instability than transistor circuits. More tube radio circuits than transistor/IC circuits are shown in this book because the book is written from a historical perspective. Converting tube circuits to transistor circuits is usually straightforward and is discussed in Chapter 5.

Keep in mind that even though all of these radio circuits work, the design shown may have been chosen to illustrate a particular point, and the design may not be optimum. For instance, in the drawings of this chapter, headphones are shown wired in the plate circuit of the vacuum tubes as was common in early receivers. A disadvantage of this is that the phone leads are at a fairly high voltage with respect to the circuit ground and touching the phones terminals could deliver a shock.

Table 1-1 shows the frequency span of the nine international shortwave broadcast bands and the lower frequency amateur bands. In the designs of this book, radios cover only one international broadcast band, or ham band, per sweep of the tuning capacitor. A different tuning coil is switched in for each shortwave band received. The ham bands are considerably narrower than the international broadcast bands.

The tuning inductor and variable capacitor are not specified in the drawings of this book because the value of these components has to be selected for the particular frequency band to be received. In most cases, the tuning capacitance, shown in the drawings as a single variable capacitor, can be realized by the combination of a 119 pF fixed capacitor in parallel with a 4 pf to 20 pF

TABLE 1-1
Frequency Ranges of the Shortwave Bands
and Band Inductance for Receivers in this Book

Band Name	Band Type	Official Frequencies	Inductance for C =150pF
160 Meters	Ham	1.81-2.0 mHz	57 µH
120 Meters	Broadcast	2.3-2.495 mHz	36 µH
90 Meters	Broadcast	3.2-3.4 mHz	19 µH
80 Meters	Ham	3.5-3.8 mHz	15 µH
75 Meters	Broadcast	3.9-4 mHz	13 µH
60 Meters	Broadcast	4.75-5.06 mHz	8.6 µH
49 Meters	Broadcast	5.95-6.2 mHz	5.6 µH
40 Meters	Ham	7.0-7.1 mHz	3.9 µH
41 Meters	Broadcast	7.1-7.3 mHz	3.9 µH
31 Meters	Broadcast	9.5-9.9 mHz	2.2 µH
30 Meters	Ham	10.1-10.15 mHz	2.2 µH
25 Meters	Broadcast	11.65-12.05 mHz	1.5 µH
21 Meters	Broadcast	13.6-13.8 mHz	1.0 µH
20 Meters	Ham	14.0-14.35 mHz	.95 µH
19 Meters	Broadcast	15.1-15.6 mHz	.82 µH
16 Meters	Broadcast	17.55-17.9 mHz	.56 µH
17 Meters	Ham	18.068-18.17 mHz	.53 µH
15 Meters	Ham	21.0-21.45 mHz	.35 µH
13 Meters	Broadcast	21.45-21.85 mHz	.33 µH
12 Meters	Ham	24.89-24.99 mHz	.20 µH
11 Meters	Broadcast	25.67-26.1 mHz	.18 µH
10 Meters	Ham	28.0-29.7 mHz	.14 µH

variable capacitor. With this choice of capacitance, Table 1-1 provides the approximate inductance values for each shortwave band. Table 1-2 contains some of the same information for the broadcast shortwave bands along with winding

TABLE 1-2
Toroidal Inductor Winding Instructions
for the 13 International Shortwave Broadcast Bands
and the RF Choke for Receivers in This Book

Band	Official Frequencies	Approx. Induct.	Approximate Toroidal Winding Instructions
120M	2.3-2.495 mHz	36 µH	86 Turns, #32 Wire, T-50-2 Core, or 23 Turns, #22 Wire, FT-50-61 Core
90 M	3.2-3.4 mHz	19 µH	62 Turns of No. 28 Wire on a T-50-2 Core, or 17 Turns of No. 20 Wire on a FT-50-61 Core
75 M	3.9-4 mHz	13 µH	52 Turns of No. 28 Wire on a T-50-2 Core, or 14 Turns of No. 18 Wire on a FT-50-61 Core
60 M	4.75-5.06 mHz	8.6 µH	41 Turns, #26 Wire, T-50-2 Core
49 M	5.95-6.2 mHz	5.6 µH	32 Turns, #26 Wire, T-50-2 Core
41 M	7.1-7.3 mHz	3.9 µH	26 Turns, #24 Wire, T-50-2 Core
31 M	9.5-9.9 mHz	2.2 µH	21 Turns, #22 Wire, T-50-6 Core
25 M	11.65-12.05 mHz	1.5 µH	16 Turns, #20 Wire, T-50-6 Core
21 M	13.6-13.8 mHz	1.0 µH	14 Turns, #18 Wire, T-50-6 Core
19 M	15.1-15.6 mHz	0.82 µH	13 Turns, #18 Wire, T-50-6 Core
16 M	17.55-17.9 mHz	0.56 µH	10 Turns, #18 Wire, T-50-6 Core
13 M	21.45-21.85 mHz	0.33 µH	9 Turns, #18 Wire, T-50-6 Core
11 M	25.67-26.1 mHz	0.18 µH	7 Turns, #18 Wire, T-50-6 Core
	RF Choke	47 µH	29 turns, #24 Wire, FT-50A-61 Core, tap 4 turns from one end

instructions for toroidal coils that can serve as tuning inductors for the receivers of this book. Many other tuning arrangements are possible.

The Crystal Radio

The crystal radio consists of a single tuned circuit with connections to an antenna and ground followed by a crystal detector and headphones as shown in Figure 1-1A. There are many different configurations of the crystal radio. The one shown is representative and works well. Figure 1-1B is a photo of a home-built crystal radio. Figure 1-1C is a photo of a commercially manufactured crystal set.

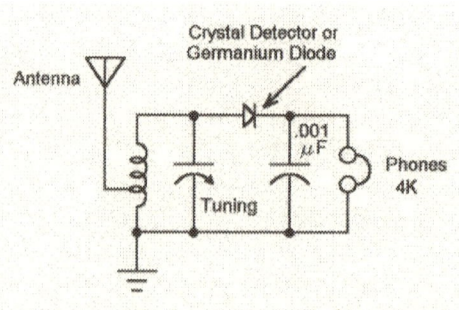

Figure 1.1A: A Crystal Radio Schematic

The crystal radio is primarily for use with medium wave AM band local stations, though shortwave reception is possible. The crystal set requires no external power. All of the components are passive. The energy necessary to power the headphones comes from the energy present in the radio wave itself. A fairly long antenna is required to extract sufficient energy from the radio wave. Crystal radio builders often construct the coils for these radios, as shown in Figure 1-1B.

It is a fascinating concept that a radio can be constructed by winding magnet wire around a wooden form and connecting a variable capacitor, crystal detector, a pair of headphones, a long wire antenna, and an earth ground connection. An AM broadcast band station can be heard in the phones at low volume, but audible!

The crystal detector was the last of several passive, or non-amplifying detectors used before the vacuum tube was invented. All of the other detectors contained motor driven parts, electro-magnets, acid baths, lamp filaments, batteries, or other aspects that were poorly suited to mobile or shipboard use and required considerable maintenance. Many of these early detectors could only detect Morse code, not voice transmissions.

The mineral crystal detector shown in Figure 1-2 requires occasional adjustment, but has none of the disadvantages of the detectors mentioned above. In Figure 1-2, A crystal substance such as lead sulfide (galena) is mounted in a metal holder which becomes one terminal of the detector. An arm with a springy wire attached is used to contact the surface of the crystal and is the other terminal

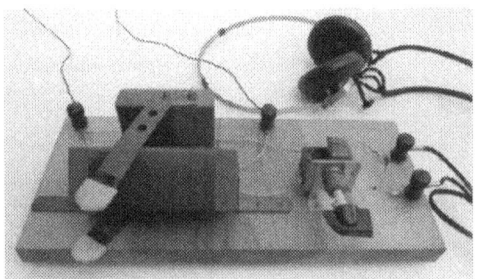

Figure 1-1B: A Home Constructed Crystal Radio

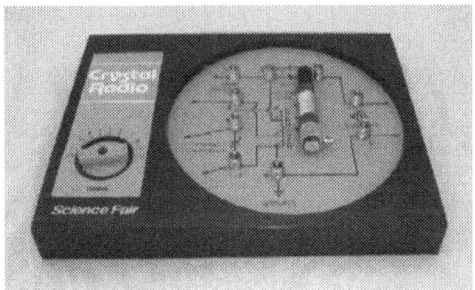

Figure 1-1C: A Manufactured Crystal Radio

of the detector. The springy wire, called a "cat's whisker", has to be moved to different locations on the crystal until a sensitive spot is found.

The operation of the crystal detector is similar to today's semiconductor diodes. A point contact germanium diode such as the 1N34 may be substituted for the crystal assembly of Figure 1-2. Germanium Diodes are pictured in the photos of Figure 1-1B and Figure 1-1C. These are much smaller than the crystal

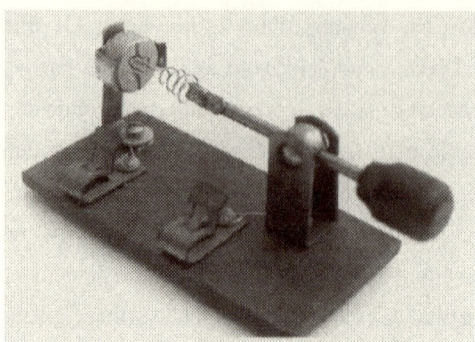
Figure 1-2: A Crystal Detector in Holder

assembly of Figure 1-2 and the diodes are not easy to see in the photographs.

Figure 1-3A shows the volt/milliampere characteristic of a crystal detector or a diode. This curve is exponential. The diode conducts in the positive voltage region of the curve and blocks current flow in the negative voltage region. Figure 1-3B shows that the diode acts like an imperfect switch, cutting off the negative part of the waveform and passing on the positive part with some distortion added.

Figure 1-4 shows the traditional switching explanation of how the diode, or crystal detector, extracts the audio from an amplitude modulated (AM) signal. There are other ways of explaining demodulation, such as thinking of the detector as a multiplying device, or by expressing the concept with mathematical equations, but the explanation in Figure 1-4 is the easiest to understand.

This explanation assumes that the diode is a perfect switch, a condition that is approached when the signal level is high (say 7 volts) and the modulation index is not too high (80% or less). The modulated AM signal in the first diagram of Figure 1-4 is followed by the same waveform with the bottom half removed by diode switching. Next, a low pass filter removes the carrier (high frequency spikes) from the waveform.

What remains is the recovered audio combined with a positive DC voltage formed by the filtered carrier. A blocking capacitor removes the DC component leaving the baseband audio signal. Headphones can, to some extent, perform the last two steps without adding other components. The reactance of headphones smooths out the high frequency carrier portion of the waveform. A small direct current combined with the audio passing through magnetic headphones does not effect the operation of the phones.

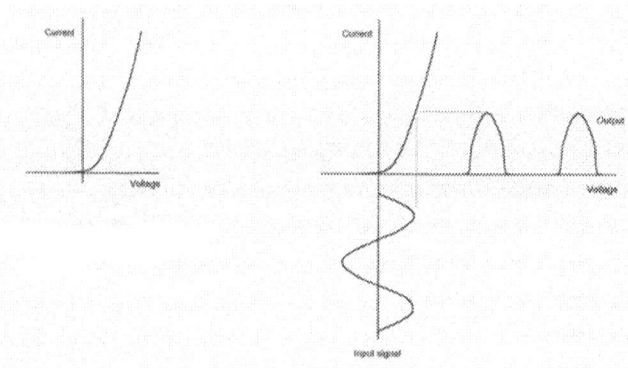

Figure 1-3A: Voltage versus Current (V-I) Characteristic of a Diode

Figure 1-3B: How the Diode Characteristic Removes the Negative Half of a Signal

Traditionally, the crystal radio was considered to be a completely passive device used with headphones. For the purposes of this book, the term "crystal radio" will be defined as any radio consisting of a single tuned circuit, an antenna and ground connection, and a non-amplifying detector that may be connected to headphones or to the input of an audio amplifier.

It was discovered in the 1950s that when the crystal set was connected to Hi-Fi amplifiers, the resulting radio reproduced high audio frequencies that were rarely heard on AM radio receivers. The high audio frequencies were heard from the crystal radio because of the wide bandwidth of single tuned circuits compared to the bandwidth of the usual superhet radio. Because of this wide bandwidth, the single tuned circuit will not always completely separate the

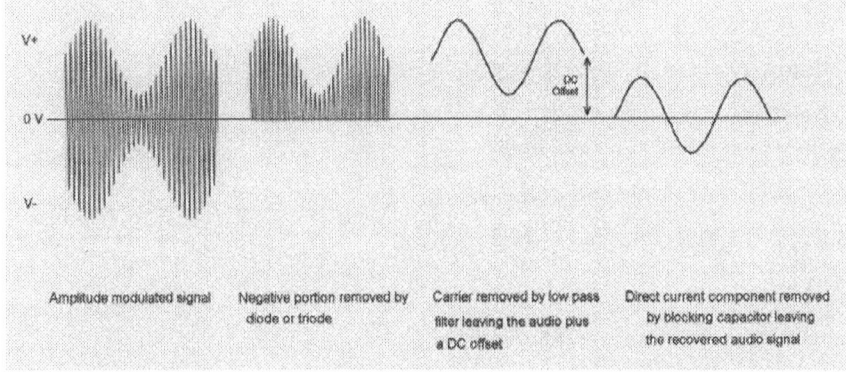

Figure 1-4: Classical Illustration of the Detection of an AM Signal

desired signal from other stations. When one station is tuned in, other stations might be heard in the background. In the 1950s, the J. W. Miller company sold a crystal radio device that was intended for use as Hi-Fi AM tuner.

Like all AM radios of the time, the crystal radio Hi-FI tuner suffered from distortion due to the non-linear V-I characteristic of the diode (Figure 1-3B). Diode distortion was greater in the crystal radio tuner because the signal was not amplified before being applied to the detector.

The author developed a crystal radio AM tuner[1] with an envelope detector instead of a diode. This design resulted in a low distortion wide bandwidth AM receiver with an unprecedented quality of reception. Most analog AM superhet radios today still use the single diode detector.

The Regenerative Radio

When the triode vacuum tube became available in 1908, the vacuum tube detector was used in place of the crystal detector of a crystal radio. The resulting circuit, shown in Figure 1-5, is called a "grid leak detector". The tube not only detected the signal, but also provided audio amplification. The 75 K-ohm plate resistor would not have been required with early battery type tubes operated with a low plate supply voltage (i.e. 22.5 volts). The plate resistor is needed with later tubes to limit the plate saturation current.

The regenerative radio is a one-tube receiver which was patented by Edwin Armstrong in 1914. A schematic of Armstrong's version of the regenerative radio is shown in Figure 1-6. The radio will work over a frequency range of 150 kHz to 30 mHz but is most

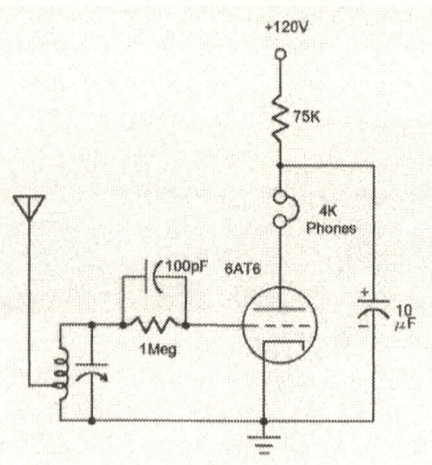

Figure 1-5: A Grid Leak Detector

---

[1] "High Quality AM From a Crystal Radio", Lyle R. Williams, *The Audio Amateur,* Issue 4-1992, pp. 24-31

useful for the shortwave, 3 to 30 mHz, frequency range. The receiver will demodulate AM, CW (continuous wave code), and SSB (single side band). SSB wasn't extensively used until the 1970s and reception of SSB is an application of the regenerative radio that was not initially anticipated.

The regenerative receiver of Figure 1-6 is essentially an RF oscillator coupled to an antenna. The circuit has variable positive feedback (regeneration) that allows the circuit to be adjusted just below the point of oscillation. This is the most sensitive adjustment point for AM reception. The gain of the circuit when almost oscillating is very high and the bandwidth is narrow.

The regeneration control is adjusted for best CW and SSB reception when the receiver just oscillates continuously. The tone of the received CW code signals can be adjusted by slight changes in the tuning capacitance. This changes the frequency of oscillation of the regenerative circuit.

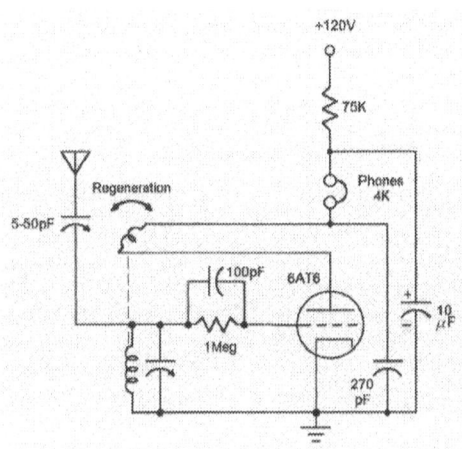

Figure 1-6: A Regenerative Radio Similar to Armstrong's Original

The regenerative circuit of Figure 1-6 is connected to the antenna through a variable coupling capacitor. Thus, during normal tuning of an AM signal, the receiver will intermittently radiate a radio signal from the antenna and may interfere with other receivers. When receiving CW or SSB, the radio will continuously radiate a signal from the antenna.

When the regenerative receiver is oscillating, the presence of the receiver is revealed to others and the location of the receiver can be determined by radio direction finding equipment. For this reason, the receiver was unsuitable for military use. The interference produced by the regenerative radio also led to its abandonment as a medium wave broadcast receiver.

The radiation from the antenna of the receiver can be greatly reduced by placing a triode buffer between the antenna and the regenerative circuit. Although the buffer provides some extra broadband gain in addition to isolation, it increases the cost of the basic regenerative radio and it has not been widely used.

It is also possible to provide antenna isolation by using a pentode tube and feeding the antenna into the screen grid instead of into the tuned circuit on the control grid. That solution to the isolation problem has also been rarely used.

Regeneration is controlled in the circuit of Figure 1-6 by mechanically rotating the tickler feedback coil inside the tuning inductor, (see Figure 7-8) thereby changing the coupling between the two coils. Intuitively, this would seem to be a good method of regeneration control; however, old publications state that this means of regeneration control changed the tuned frequency of the radio and adjustment of the control was critical or abrupt.

From about 1930, the regenerative radio has been used primarily for shortwave (3 to 30 mHz) signals with AM or CW code modulation to receive international broadcasts, amateur, and commercial communications. The regenerative radio was the only affordable receiver for most early hams. Because of the simplicity of the design, the receiver has been a favorite for home construction projects and kits. There have been very few commercially manufactured regenerative receivers in recent years. The scarcity of commercial versions may be partly due to the difficulty of obtaining FCC type approval for a radio that is prone to cause interference with other receivers.

The regenerative radio has been used mostly with a long wire outdoor antenna. When an amplifying antenna buffer is used, a short indoor antenna will often suffice for shortwave reception. The radio does not readily demodulate FM, and the regenerative radio is not very useful above 30 mHz. A variation of the regenerative radio, called the super-regenerative radio, can be used above 30 mHz and that receiver will be discussed later in this chapter. The regenerative radio can be used on medium wave and long wave, but a fixed bandwidth receiver is more convenient at these frequencies.

The TRF Receiver

When the regenerative receiver fell out of favor as a broadcast radio in the 1920s, it was replaced by the tuned radio frequency (TRF) receiver. These battery operated receivers were often called "neutrodynes" after a method of preventing instability developed by L. A. Hazeltine. Not all of the radios called neutrodynes actually used the patented neutrodyne technique. The early TRF's consisted of one or more RF vacuum tube stages followed by a grid leak detector stage (Figure 1-7). The three tuned circuits each had separate variable capacitors with dials. The three dials had to be individually set to tune in a station.

With three tuned circuits and no shielding, these radios were prone to instability and oscillation. While oscillation is necessary for the regenerative

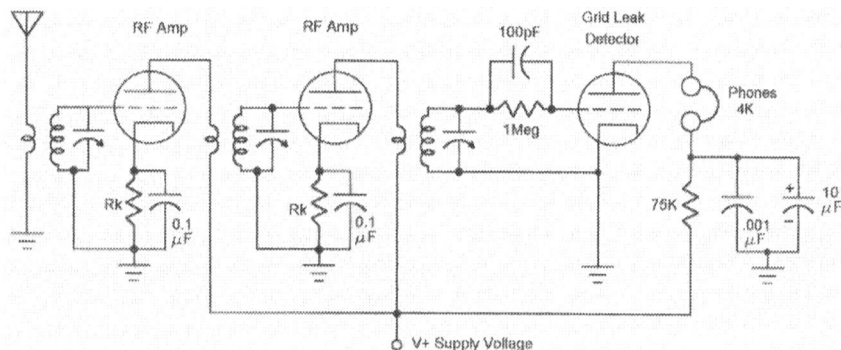

Figure 1-7: A Tuned Radio Frequency (TRF) Receiver

radio to operate, oscillation is undesirable in TRF receivers. Magnetic coupling between the three coils and electrostatic coupling between the tubes would contribute to the instability of the radio, but the major cause of instability was the plate to grid capacitance of tubes. An oscillator circuit analogous to the TRF amplifier is known as a "tuned grid, tuned plate" oscillator.

The magnetic coupling between coils could be minimized by positioning the inductors at certain angles with respect to each other. Various techniques were tried to counter the grid-plate capacitance of the triode tubes and prevent oscillation. The best technique was to "neutralize" the grid plate capacitance by placing a small capacitor in a feedback circuit through which an out-of-phase signal was passed. The grid-plate capacitance of the tube was counteracted by the added feedback.

Later, pentode tubes were introduced that were stable at medium wave frequencies without neutralization. Metal shielding of the coils and tubes was introduced to stop coupling between them. Speakers consisted of a large horn attached to a headphone like element. The appearance of these horn speakers was similar to the mechanical phonograph horns of the day.

The speaker had about the same impedance as headphones and could be driven by the same audio tube. AC operated TRFs that contained a single ganged variable capacitor and an electrodynamic type speaker became available in the late 1920s. Stations were tuned in by turning a single knob.

The TRF receiver is primarily useful as a medium wave set. The receiver lacks sufficient selectivity for shortwave. The bandwidth of the receiver is wider at the high frequency end of the dial. A detailed account of the design of the TRF receiver is covered in a book by David Rutland[2].

The Superheterodyne Receiver

Shortly after Edwin Armstrong invented the regenerative radio, he started working on the superheterodyne receiver. A block diagram of the superhet is shown in Figure 1-8A. One of many possible implementations of the superhet

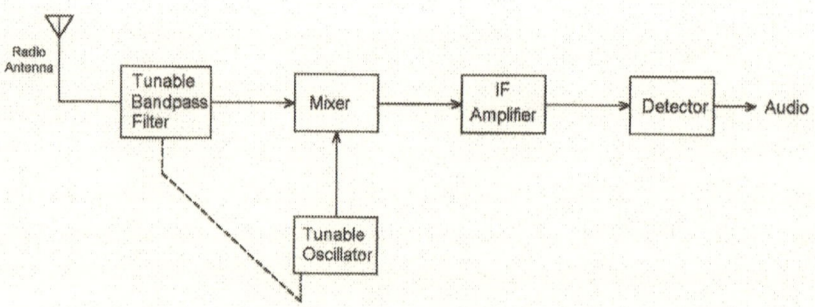

Figure 1-8A: Block Diagram of a Superheterodyne Radio

radio is shown in Figure 1-8B.

When two RF signals are electronically multiplied by each other, the output is two other signals. The frequency of one of the output signals is equal to the sum of the frequencies of the original two signals and the other output signal is the difference of the original two frequencies. Suppose that one signal is the desired radio station whose frequency is the Fs (signal). The other signal is the local oscillator of the receiver and has the frequency Fl. When the two are electronically multiplied (mixed), the result is two signals of frequency Fif (intermediate frequency) and Fim (image frequency). Assuming that Fl is larger

---

[2] *Behind the Front Panel,* David Rutland, Wren Publishers, Pilomath OR, 1994

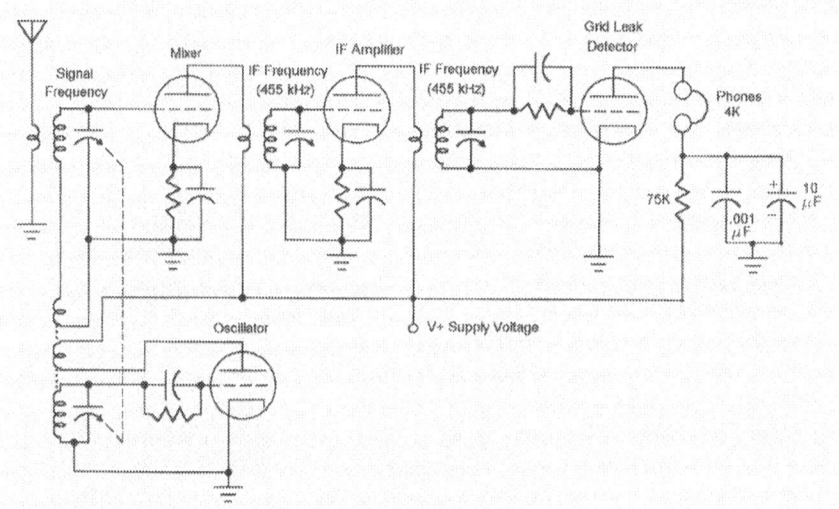

Figure 1-8B: Schematic Diagram of a Superheterodyne Receiver

than Fs, the frequencies Fif and Fim are as follows:

$$Fif = Fl - Fs \quad Fim = Fl + Fs$$

The intermediate frequency, Fif, is the desired result; the image, Fim, is undesired and must be filtered from the output. When tuning in stations, the local oscillator frequency, Fl, is changed so that the output, Fif, will always be the same frequency, the intermediate frequency (IF). (Note that the variables in the above equations represent the frequencies of radio signals, not the amplitude).

Since the intermediate frequency remains the same, the intermediate frequency amplifier can be tuned to one frequency only (the IF) and left there. Typical IF frequencies are 455 kHz for medium wave AM, and 10.7 mHz for VHF FM broadcast radios. Usually, the IF is lower than the received signal. A more sensitive and narrower band amplifier can be built for the IF frequency than for the received signal frequency. Also, the bandwidth of the receiver will remain uniform regardless of the received frequency.

Armstrong's original reason for building the superhet was to receive signals of a higher frequency than was otherwise possible at that time. Once the higher signal frequencies were converted to the lower IF frequency, the signal could be amplified at the high gain and narrow bandwidth possible at the IF frequency.

Note from Figure 1-8A that the input filter must remove the image frequency or a given signal will show up at two different locations on the tuning dial. In medium wave receivers, the tuning shaft of the input filter is coupled to the tuning shaft of the local oscillator in order to reject the image frequency and enhance the gain at the desired frequency. The input filter and the local oscillator must "track" as the receiver is tuned. Tracking means that the input bandpass filter must pass a signal frequency that when converted by the local oscillator and mixer, will equal the intermediate frequency. Tracking between the input tuned circuit and the local oscillator is one of the more difficult problems of superheterodyne receiver design.

Communication receivers often use dual conversion to improve the image rejection. In dual conversion, the signal frequency is first converted to a high IF frequency and then converted back to a low IF frequency such as 455 kHz. Suppose we wish to receive a signal of 10 mHz where the intermediate frequency is 455 kHz. Then the local oscillator frequency, Fl, is 10.455 mHz, (Fs + Fif) and the image is 10.91 mHz (Fl + Fif). The image, 10.91 mHz, is so close to the received signal, 10 mHz, that it is difficult to remove the image frequency from the signal presented to the mixer.

Suppose we make the IF frequency 15 mHz instead of 455 kHz; then if the signal is 10 mHz, the local oscillator frequency is 25 mHz. The image will be at 35 mHz. Here the image is 3.5 times the signal frequency and can easily be filtered out. The 15 mHz IF frequency can then be converted to a second IF frequency of 455 kHz for easy amplification and bandwidth control. If 15 mHz IF frequency was within the signal frequency range that is to be received, the receiver would pick up its own IF frequency and would possibly be unstable. So conversion frequencies have to be carefully chosen to work with the frequency range that the radio is to receive.

About 1930, the superhet replaced the TRF radio as the dominant medium wave design. Almost all receivers today use the superheterodyne principle. That includes AM and FM broadcast radios, televisions, satellite receivers, aircraft receivers, radar, cellular phones, wireless local area networks (LANs), and nearly every other type of radio receiver. Digital receivers are also usually superhets.

The Super-Regenerative Receiver

The super-regenerative receiver will receive at the VHF and UHF frequencies above 30 mHz. The receiver is simple, can be constructed quickly, and is something of a novelty. If powered by batteries and listened to on headphones, the receiver can be implemented with a single tube or transistor. A representative schematic is shown in Figure 1-9A. A regenerative circuit is cycled into and out of oscillation at a cycle rate in the supersonic range, such as 30 kHz, as shown in Figure 1-9B.

The output of the receiver before detection and filtering is a train of RF pulses at a pulse rate called the quench frequency. The width of the RF pulses varies according to the audio level of the signal. When this width modulated pulse train is detected and filtered with a low pass filter, the audio signal from the radio wave is recovered. The frequency of the quenching operation in a self-quenched super-regen is controlled by the grid resistor and coupling capacitor. The quench frequency will change as the signal frequency changes. The quench frequency can be kept constant by using a separate quench oscillator.

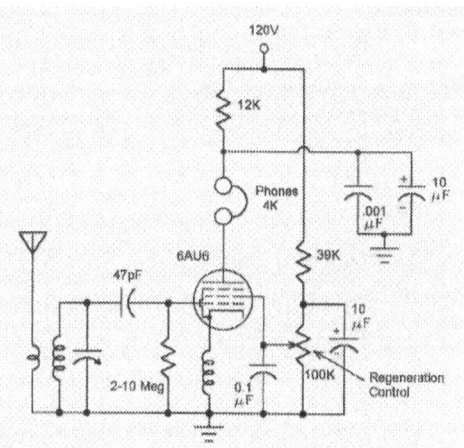

Figure 1-9A: A Super-regenerative Receiver

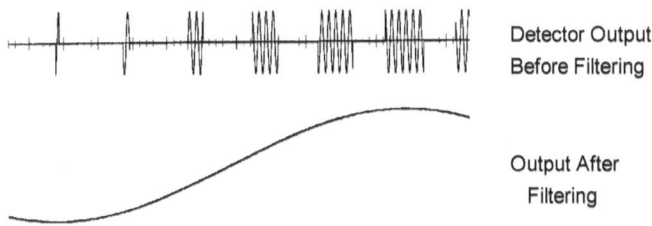

Detector Output Before Filtering

Output After Filtering

Figure 1-9B: Output Waveforms of a Super-regenerative Receiver

The super-regenerative receiver is both impressive and disappointing. For a one tube or transistor receiver it provides performance that would require several tubes in a superhet design. On the other hand, the bandwidth of the receiver is several tens of kilohertz, which is too wide for optimum amateur radio and communications uses. The sensitivity of the receiver is about 10 microvolts. The signal to noise ratio (SNR) is poor and the radio resists attempts to improve any of these characteristics. Adding an RF amplifier in front of the super-regenerative stage does not significantly improve the sensitivity or SNR of the receiver.

The super-regenerative receiver was popular in the 1930s for VHF reception. It was used by hams as late as the 1950s, for cheap citizens band (CB) receivers in the 1960s, and in the late 1990s as a data receiver for local wireless computer networking systems.

Due to the wide bandwidth, the super-regenerative receiver works as an FM broadcast receiver and as a receiver for the sound portion of television signals. Technically, the super-regenerative receiver demodulates AM only. But by using the edge of the tuning curve of the receiver to convert the FM to AM, the receiver can demodulate FM fairly well. The super-regen cannot decode the stereo subcarrier on FM broadcasts or on television sound signals. Furthermore, the stereo subcarrier beats against the quench oscillator of the super-regenerative radio, producing weird audio artifacts.

With many cheap FM radios such as the "Walkman" on the market, it would not make economic sense to construct a super-regenerative radio for FM broadcast reception. However, this might make a good project for children.

A possible use of the super-regen would be as a portable TV sound receiver. It could be used to listen to local TV news programs at various locations when it is not necessary to see the television picture. There are commercial portable TV sound receivers available for the VHF TV channels, but none for the UHF or cable TV channels. The super-regenerative receiver is so simple and cheap that a separate radio could be built for each desired TV channel.

The Direct Conversion Receiver

The direct conversion receiver is a special case of the superheterodyne receiver in which the local oscillator frequency is the same as the signal frequency. This results in an intermediate frequency of zero Hz. A block diagram is shown in Figure 1-10. No schematic is provided here, but details can be found in any of the later issues of the *ARRL Handbook* such as the 1988 issue. As this is a fairly recent receiver design, only transistor versions of the circuit are likely to be found.

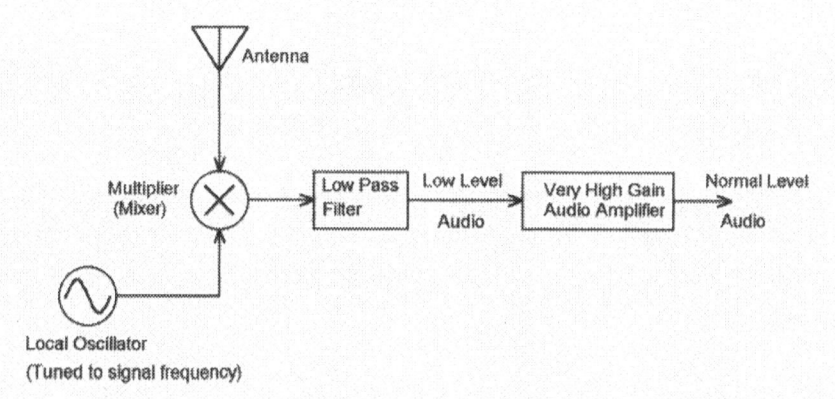

Figure 1-10 Block Diagram of a Direct Conversion Receiver

The output of the mixer in Figure 1-10 is a direct current (DC) signal (the detected carrier) which is combined with the demodulated audio (the detected sidebands). Since the local oscillator signal is the same frequency as the signal frequency, the image frequency is at two times the signal frequency and no image filter other than the audio low pass filter is required. The receiver detects SSB and Morse code (CW) signals without using a diode or grid leak detector and without needing a separate beat frequency oscillator (BFO) as required in superhet receivers.

Direct conversion receivers are not intended for AM reception. When receiving an AM signal, the carrier will beat against the local oscillator. Unless the local oscillator is tuned to the exact carrier frequency, an undesired audio tone will be produced.

There is no RF gain before mixing and the audio output is at a very low voltage level, equal to the level of the signal on the antenna. An audio amplifier with an unusually high gain is required following the mixer. High gain audio amplifiers tend to produce a lot of hum and noise.

The receiver tunes SSB in a similar manner as a superhet with BFO. When the local oscillator frequency exactly equals the suppressed carrier frequency, the audio output will be undistorted. A slight detuning of the local oscillator will produce distorted, but intelligible audio.

CW is received by intentionally slightly detuning from the signal frequency so that the code signal will have an audible audio tone. Direct conversion receivers have been built and used by amateurs. The direct conversion concept is currently being used in cellular telephone handset designs.

Conclusion

The six types of receivers that have just been discussed are the only receiver configurations that have yet been developed. With the exception of the direct conversion receiver, all of these receiver configurations were in use by 1930. The advent of digital radio has not added to the list. Most digital radios are superhets.

2 Types of Digital Receivers

Digitally Synthesized Analog Superhet

The term digital radio has several meanings. In the digitally synthesized superhet (Figure 2-1) the local oscillator frequency is synthesized by dividing the signal from a crystal controlled oscillator using digital frequency dividers. The frequency received is selected by typing it in on a keypad, using an up/down switch, or a rotary encoder wheel. Most automobile radios and commercial shortwave receivers today use this technique.

Tuning of the synthesized receivers is very precise, but only discrete frequencies can be accessed. If a medium wave receiver tunes in 10 kHz steps, then it is not possible to tune exactly 1544 kHz; 1540 kHz or 1550 kHz must be selected instead of the desired frequency. Automobile radios usually don't have a keypad or rotary encoder for frequency selection. Stations are selected by scanning or by single step push button up/down tuning.

Figure 2-1: A Digitally Synthesized Analog Receiver

Some shortwave receivers will step in frequencies as small as 100 Hz. With a fixed beat frequency oscillator (BFO) signal, this allows SSB signals to be tuned with reasonable, but not perfect accuracy.

New frequencies on synthesized receivers are found by scanning a desired band. Receivers that have a rotary encoder wheel will tune up or down in frequency in a way similar to tuning an analog receiver. To tune to an exact frequency such as 5975 kHz, punch in those numbers on the key pad and then press enter. To find a station between 5940 and 6200 kHz, use the scan function between these frequencies. The radio will scan until it finds the first signal above the threshold of the scanner.

The digital scanner will miss stations that are below the signal threshold of the scanner. The digital synthesis radio usually takes longer to search for new stations than would be required using an analog receiver. Both digitally synthesized and conventional analog superhets have their advantages. Digital synthesis receivers have been manufactured since the mid 1980s.

All-digital Receivers

In all-digital radios, sometimes called software receivers, the antenna is followed by an analog bandpass filter connected to an analog to digital (A/D) converter. The A/D converter samples the signal from the antenna and outputs a string of numbers representing the signal. From here on, the radio resembles a computer. All of the usual radio functions such as conversion, filtering, amplification, detection, etc., are implemented by software in the digital domain. The software contains mathematical formulas that operate on the string of numbers. The mathematical equations that were developed to predict the behavior of analog circuits are programmed into software and used directly to become the actual radio. The type of radio (superhet, TRF, etc.) will be determined by the software. The superhet is the radio type most used.

The all digital radio can decode AM or FM signals or any other type of modulation. At the output of the receiver, the audio represented by numbers must be converted back into an analog signal by a digital to analog (D/A) converter. The analog signal is fed to a conventional power amplifier and speaker. All digital consumer radios are not yet widely available, but the concept is used in the design of cellular telephones.

Digital Modulation

Digital Modulation means that the audio is converted to a string of numbers by an A/D converter. The numbers are processed, compressed and transmitted over the air. The modulation method is one of the digital types such as quadrature frequency shift keying (QFSK). The receiver has to recover the string of numbers and convert them back into an analog signal.

This technique is currently used by the two satellite broadcasting services, XM and Sirius. Since the signal comes from overhead satellites, signal dropouts are rare except when under a metal structure. When a dropout does occur, it takes around five seconds for the radio to recapture the signal.

The FCC is planning to change the modulation technique on both the medium wave AM band and on the VHF FM band to digital. The reason for this is that the quality of reception on the 10 kHz wide AM bands can be enhanced to equal that of current FM stereo, and the quality of reception on the 200 kHz wide channels of the VHF FM band can be increased to that of current audio CDs. It has been suggested that AM frequencies will regain some lost popularity because the quality of reception will be good, reception will be in stereo, and medium wave radio stations are cheaper to build than VHF radio stations.

The emergence of terrestrial digital broadcasting in the U.S. has begun in a few areas as of this writing. Present systems transmit both analog and digital signals. When the digital signal drops out, the system reverts to the analog signal. The analog signal will be used by the great majority of listeners who have not yet purchased a digital receiver.

Audio quality on current domestic analog AM and FM stations is already quite good in strong signal areas and analog fringe area reception is more usable than digital. If total conversion to digitally modulated terrestrial radio is made and the analog part of the signal is not retained, countless millions of analog AM and FM radio receivers will become unusable. Most regrettable, is that antique AM and FM receivers will no longer work. Digital receivers are not a good hobby project because of the extreme complexity of the circuits and the expensive design equipment required. As of this writing, terrestrial digital receivers are much more expensive than the equivalent analog receiver.

Digital broadcasting can achieve a larger coverage area without increasing transmitter power over analog levels. Since AM and FM radio stations are allotted a limited distance of coverage, the transmitted power for the same distance coverage will be greatly reduced. The advantage will be to the radio station owners who will have lower electric utility bills.

Analog AM signals in fringe areas fade gradually until the signal is below the noise level, below the level of another station on the same frequency, or below the threshold level of the receiver. A transmitted analog voice signal can often be understood even when the signal is buried in noise and congestion.

In a fringe area, the reception of digital radio signals is practically useless. When the digital signal of a totally digital station drops out, the output of the receiver reverts to complete silence and stays silent until the receiver can recapture and load the signal. This renders fringe area speech very difficult to understand.

This is the same effect encountered when digital cellular phones are operated in a low signal area. The effect of digital signal dropout on music reception is equally frustrating. Receiver dropouts will be more troublesome with terrestrial digital broadcasting than with satellite broadcasting because of interfering objects between the transmitter and receiver. It is not clear what the effect on reception of digital modulation signals from two stations on the same frequency will be.

There is no move at present to convert shortwave broadcasts to digital. Some international broadcasters, however, have invested in research into digital broadcasting techniques. The quality of shortwave reception could benefit considerably from digital modulation techniques if the problem of dropouts could be solved.

But the audience for shortwave broadcasting includes non-affluent third world listeners, who would not be able to purchase new digital receivers in the foreseeable future. As with local radio, the goal of shortwave radio stations is to reach the maximum number of listeners possible. Therefore, I suspect that shortwave receiver builders will have access to analog signals for an indefinite period of time.

Spread Spectrum

Strictly speaking, spread spectrum was developed before the advent of digital electronics. However, spread spectrum was not very practical until digital circuits became available. Spread spectrum is a technique of spreading a radio signal over a wide bandwidth of about one megacycle. The transmission power can be so low that the received signal is below the atmospheric noise level. An ordinary narrow band receiver would not even indicate the presence of a spread spectrum signal. But the spread spectrum receiver can recover the signal out of random noise. Several spread spectrum signals can be transmitted over the same frequency space. Obviously, at some point adding extra signals in the same bandwidth will result in reception problems. It is not known how many signals can reliably occupy the same frequency space.

The frequency hopping type of spread spectrum transmission was patented in the U.S. during World War II in the name of the female film actress, Hedi Lamar. It is not clear whether the spread spectrum idea was that of Lamar herself, or an associate. However, Lamar has been inducted into the Engineering Hall of Fame because of her patent.

Spread spectrum has been used since the 1980s for army battlefield communications. The modulation process is similar to encryption techniques. For the signal to be recovered, the receiver must contain information of how the transmitter is modulating the signal. Therefore, spread spectrum is inherently a high security medium even without additional encrypting of the signal. Interception by unauthorized receivers is very difficult. More recently, the spread spectrum technique has been used in ordinary portable telephones and in wireless computer networks.

There are two types of spread spectrum, frequency hopping and chip sequence encoding. With frequency hopping, the transmitter carrier frequency is semi-randomly changed every so many milliseconds. The modulated transmitter carrier signal executes a sequence of frequency hops within the assigned band space. The receiver must be programmed with the frequency hopping sequence so that it can follow the transmitter through its frequency excursions.

In another spread spectrum technique called the chipping code method, the signal to be transmitted is mixed with a digital chipping signal. This results in a wide band signal for transmission. The receiver must have the chipping code in order to recover the original signal.

Spread spectrum is best suited for short range highly secure communication. It has not yet been used for broadcasts.

3 Radio Services Currently Available

When I was growing up in the 1940s, the prominent entertainment technologies were radio, phonograph, and motion pictures. In the 1950s, television and the HiFi/Stereo revolution in recording and reproduction of music emerged. In my view, television was a nice, but unnecessary enhancement to radio. HiFi reproduction was, and still is, of considerable interest. To me, radio is the most fascinating invention of all time. People who like to build radio receivers are naturally interested in what can be picked up on those receivers. The purpose of this chapter is to provide a discussion of currently available radio services. Some of the services discussed here cannot be received on homebuilt radios. A perspective is provided from which the radio receiver building and listening hobby can be viewed.

Long Wave

The long wave band is the region from 10 kHz to 520 kHz. Modulation is AM, Morse code, and digital. The band is used for commercial broadcasts in Europe, but in the U.S. it is used for beacons, navigation signals, aviation weather, the WWV time station and similar services. Long wave is more subject to atmospheric and manmade noise than any other band. Receivers for this band require a very long antenna or an active antenna for good reception. Loop antennas are useful and they make possible the separation of stations on the same frequency according to the direction of the station from the receiver. The crystal set, the TRF, and the regenerative receiver can be used on long wave, but these receivers are not ideal for the purpose. The superhet radio is most commonly used on long wave. Many radio listeners find that there is a lack of interesting signal content on this band. Some, however, make a hobby of finding and identifying long wave signals.

Medium Wave

The medium wave band, usually called the "AM band" contains U.S. commercial AM broadcast stations and this radio band once held the prominence enjoyed by television services today. The frequency span is from 530 to 1710 kHz. AM radio was the only home entertainment source other than the phonograph in the 1940s. There were national radio networks supplying programming to affiliated stations as is the case with network television today.

Medium wave acts as two different bands depending on whether one listens during the daytime or at night. During daytime, the distance of reception is limited to about 100 miles, as is the case with FM stations both day and night. At night, reception of stations thousands of miles away on medium wave is possible. But the FCC assigns many U.S. stations to operate on the same frequencies; thus, the long distance capability of nighttime medium wave stations is not realized. Most AM frequencies at night contain a roaring muddle of stations that is useless to the radio listener. A loop antenna may be of some help in separating stations on the same frequency.

There are a few frequencies that are called clear channels and are supposedly assigned to only one U.S. station. These clear channel stations are allowed to operate at higher power (50,000 watts) than other medium wave stations. Theoretically, a clear channel station can be heard over most of the U.S.. For the last 20 years, however, the FCC has assigned other low power stations to the clear channel frequencies. The reception at night of clear channel stations is usually poor. In the past, clear channel stations provided programming of interest to wide audiences, somewhat like network programming. Today, clear channel programming is primarily of local interest.

On the AM broadcast band, much as on the FM broadcast band, most areas of the U.S. provide programming of interest to a limited audience. In much of America, the AM band is the home mostly of country and western music, one or more kinds of rock music, conservative talk shows, sports, and countercurrent Christian programming. Surprisingly, a few medium wave stations in low population areas provide programming such as 24-hour news, jazz, or oldies music. The widest range of radio broadcasting content is found in large cities.

About a decade ago, the FCC extended the top end of the AM band from

1610 kHz to 1710 kHz. The new stations in this frequency range are of the same type as on the rest of the medium wave band.

Like long wave, medium wave is subject to considerable atmospheric and manmade noise. The audio quality of local medium wave stations can be good, though not as good as FM. Except in stormy weather, the AM band is capable of good quality music reception with or without stereo. By building a special receiver, mono AM reception of local AM music stations, where they exist, can be greatly improved (see Footnote 1 of Chapter 1).

Stereo broadcasting significantly improves the sound of AM radio. Stereo broadcasts were prohibited on AM in the 1970s by the FCC in order to increase the popularity of FM stations, which were allowed to broadcast in stereo. Later, when AM stereo was permitted, the FCC declined for several years to standardize one of the three AM stereo systems then available. As a result of these FCC policies, most U.S. AM stations today do not broadcast in stereo and most listeners do not have AM stereo receivers.

The current state of decay of domestic AM medium wave programming is unfortunate. The situation on AM in most areas of the country leaves little of interest to listen to on antique radios that receive only this band. Sometimes antique AM broadcast receivers are used with a shortwave converter to receive the shortwave broadcast bands. Some antique radios already have shortwave bands, thus the converter is unnecessary.

Whatever the programming, listening to AM stations on a crystal radio over headphones can be a fascinating experience. With no amplification, the sound level is very low, but the stations are exceedingly crisp and clear. In addition to the crystal set, regenerative, and TRF radios can be used on AM but superhets are most common.

Shortwave

On shortwave, stations from all over the world can be received at any location. There are thirteen international broadcast bands on shortwave between 1.7 mHz and 30 mHz, as shown in Table 1-1 of Chapter 1. For some, shortwave listening is a hobby in itself. Shortwave stations can be readily received on homebuilt receivers. The shortwave band contains the same countercurrent

Christian and political programming as does the AM broadcast band, but there is much more. Shortwave is a primary source of world news. A majority of the stations are in languages other than English. However, there are more English stations on shortwave than stations of any other language.

Countries where English is not the primary language often have English language broadcasts. The government stations of Belgium, The Netherlands, France, Germany, Sweden, Switzerland, and many other countries have programming in English. The programming provided by these European countries is often informative and refreshing. The news from these stations is unbiased by corporate interests and tends to avoid the government influenced reporting and the emphasis on entertainment and celebrities that dominates U.S. news. Even the U.S. government shortwave station, the Voice of America (VOA), provides better news than U.S. corporate stations.

Shortwave contains exotic music from different parts of the world that is unlikely to be heard elsewhere. The sound quality on shortwave ranges from good to very poor. Multi-path, ionosphere reflection, fading, and congestion conspire to create many kinds of audio distortion. The best quality of reception is usually on the higher frequency bands (15 mHz and up) during local daytime. Strong stations on lower frequencies can be received with adequate quality for music listening at night. Although stereo is possible on shortwave, it has not been used except for occasional testing.

Shortwave stations often transmit simultaneously on multiple frequencies and change their frequencies frequently to take advantage of changing propagation conditions. Finding and staying with a station can be somewhat of a challenge. Also, stations may stay on the same frequency, but switch languages from time to time.

The regenerative radio can deliver remarkably good sound from medium strength signals on shortwave. For stations that transmit mostly speech, the listener is usually satisfied when it is possible to "copy" or understand the spoken words.

Shortwave contains the amateur HF bands. A large number of operating modes are used. AM was common in the past, but CW and SSB are most common today. The regenerative radio was the receiver used mostly by amateurs in the early days of radio. Only wealthier hams could afford the communication superhets offered by Hallecrafters, National, and Hammarlund.

Modulation on shortwave broadcast stations is mostly AM. The non-broadcast shortwave bands contain every type of modulation imaginable. Single sideband suppressed carrier (SSB) AM is a favorite form of modulation for two way voice communications such as ham radio. SSB greatly reduces the amount of power necessary to transmit a voice signal.

In order to receive SSB, the suppressed carrier must be reintroduced by a local oscillator in the receiver called the beat frequency oscillator (BFO). The BFO frequency must be fairly close to the suppressed carrier or the speech will be unintelligible. If the BFO frequency is close to, but not right on, the correct frequency, the speech will be intelligible but distorted. The BFO cannot be kept at exactly the right frequency for undistorted speech without frequent readjustment.

SSB is used by a small number of shortwave broadcast stations such as HCJB in Ecuador. The reception of speech from these stations is acceptable, but the reception of music is awful.

The receivers that are used on shortwave are the superhet, the regenerative, and direct conversion types. The most powerful shortwave stations on frequencies of five to seven mHz can be received with the unpowered crystal set.

The FM Broadcast Band

The commercial frequency modulation (FM) broadcast band is in the frequency band at 88 to 108 mHz. This band is between the frequencies of VHF television channels six and seven. Atmospheric and manmade noise at these frequencies is significantly lower than on the medium wave and long wave bands. The maximum distance range of FM stations is about 100 miles both day and night. VHF frequencies are limited in range by the line of sight on the Earth's curved surface. Thus, at greater than 100 miles, the signal leaves the Earth and goes off into space. As on the AM broadcast band, the FCC assigns stations about 200 miles apart to the same frequencies. This causes problems when trying to receive far away FM stations. However, interference between FM stations on the same frequency is no worse at night than during the daytime.

Several times a year, there are rare atmospheric conditions called "E-skip" under which VHF signals are reflected from the "E" layer of the ionosphere and FM stations from thousands of miles away can be heard for a half an hour or so usually around sunset or sunrise. E-skip is fun to observe, but the effect is too infrequent and too short lived to be of practical value.

The FM band provides lower noise, a higher audio frequency capability, and most, but not all, stations broadcast in stereo. Lower noise is achieved partly due to the lower atmospheric noise at VHF and to the fact that FM receivers reject amplitude variations. Noise is primarily amplitude modulated. FM stations occupy a bandwidth (200 kHz) of 20 times the bandwidth of AM Stations (10 kHz). Thus FM can be thought of as a form of the spread spectrum technique that uses wider bandwidth to achieve less interference. Spread spectrum was discussed in Chapter 2.

The FM bandwidth contains more music stations than talk stations since reception is usually better than on AM. As with AM, programming on FM is limited to a rather narrow range of interests with the widest range found in large cities. The portion of the FM band between 88.1 and 91.9 mHz is reserved for non-profit stations. The National Public Radio (NPR) stations, operated on college campuses, are found in this section of the FM band. Christian stations are found mostly on the non-profit FM frequencies, but some are in the commercial part of the FM band as well.

Building home receivers for FM is possible, but infrequently done. An FM superhet is more complex and requires higher frequency components than the medium wave counterpart. Commercial FM receivers such as the "Walkman" variety are widely available and cheap.

One VHF receiver which is exceedingly simple is the super-regenerative radio discussed in Chapter 1. The super-regenerative radio can operate throughout the VHF and UHF frequency regions, but the performance is very limited. This receiver can be built with a single transistor or tube, but the quality of reception is poor, noise is high, and the radio is adversely effected by a stereo subcarrier in the FM signal.

Above 108 mHz

Voice modulation above 108 mHz is primarily FM. Television audio is wide band FM and communications transmissions are in narrow band FM. One exception is the VHF aircraft band, which uses AM transmission because the quality of reception of FM signals would be degraded by the Doppler effect of moving airplanes.

The frequencies above 108 mHz are almost exclusively the realm of the superhet radio. The vast majority of radio signals are now on these higher frequencies. This includes most of broadcast television, cellular phones, aircraft communications, satellite communication, portable phones, wireless networks, walkie talkies, radar, police, military, space, terrestrial microwave links, ham radio, and numerous other services.

Internet Radio

Many radio stations, both local and foreign, can be listened to on the Internet. This offers an alternative to listening to the stations with a receiver. Radio stations can be heard that are out of the listening range of a radio receiver. Different sound compression and encoding schemes are used, but the most popular is Real Audio. Often, two qualities of compression are used. The lower quality compression is used with slow telephone line connections to the Internet. The low quality sound can be streamed with fewer dropouts while the download catches up, but the sound is worse than medium wave over the air radio reception. The higher quality compression format can be used with fast connections to the Internet. The quality is currently about as good as FM stereo reception and will probably improve as the technology progresses. An individual song or five minute speech segment can be downloaded to the computer's hard drive and then played from the hard drive without interruption.

The disadvantages of Internet radio are listed here:

1. A computer has to be utilized to listen to the Internet stations. It is unlikely, but not impossible, for the computer to be used for other tasks while downloading and playing audio.
2. A phone line or other Internet connection are tied up while listening to radio. Other Internet projects cannot be performed at the same time.
3. Phone line connections to Internet are usually too slow for the best quality

Internet radio listening.
4. The required Internet connection means that the system is not portable even if a laptop computer is used. A wireless laptop Internet connection could be used, but away from home this service is only available in certain restaurants, hotels, and other business locations and is not always free.
5. The audio quality of Internet radio currently ranges from poor to good compared to the quality of satellite radio reception which is excellent.

At present, Internet radio is no substitute for broadcast radio. Nevertheless, there is good programming on Internet radio and it makes possible hearing distant AM and FM radio stations that cannot be heard any other way. It can be useful when the computer and Internet connection are not required for other things. Though sound quality is limited, there is no noise, fading, multipath, or interference from other stations.

Big Dish Satellite Radio

The programming of many radio stations, including shortwave stations, are available from satellites. This is another alternative to listening to radio with a conventional receiver. The shortwave stations often announce the satellite names and transponder numbers on their over the air broadcasts. To receive these stations, one needs a large dish satellite receiver system, the same that is used for free satellite television and which costs on the order of $2000.

Changing between two radio stations might involve moving the dish from one satellite to another. Since there may be hundreds of radio and TV channels, finding a given radio station on a satellite can be challenging.

Sound quality on satellite is superb. For those who can afford it, and have the patience for tuning big dish satellite radio stations, this is a very good source of programming. A large number of TV channels and network TV feeds can be received using the same large dish equipment. Due to the high cost and non-portability of big dish satellite radio, it is not a substitute for a radio receiver.

It is not clear how digital TV broadcasting will affect big dish TV. Digital signals can be easily scrambled. All domestic TV signals will ultimately be changed to digital. Most TV services today don't mind if their signal is intercepted via satellite. These signals are ultimately broadcasted over the air anyway. The exception is pay for view TV channels, and these already scramble their analog satellite signals.

XM and Sirius Satellite Radio

Perhaps the ultimate radio experience is provided by the two new satellite radio services, XM and Sirius, which offer high quality reception most anywhere in the U.S. in cars or at home. The antenna required is small and compact. Both services provide over 100 radio channels for cars and homes via satellite transmission. There are receivers that can be used both in an automobile and at home. The subscription fee is around $13 a month for a cutomer's first receiver. The satellite receivers must be purchased by the user and automobile receivers are currently priced at $250 and up. Some new cars can be purchased with a satellite radio receiver installed. With over 100 radio channels per service that can be received anywhere in the U.S., satellite radio finally provides news and music for nearly everyone.

Summary

FM stations are local only, the audio quality is very good for stations that are in range, and stereo broadcasting is almost universal. Available programming in most areas is of limited interest.

If a local AM music station of interest exists, it is possible to build a mono receiver that is superior in quality to any AM radio you are likely to be able to buy commercially (see Footnote 1, Chapter 2). AM stereo has been all but abandoned in the U.S. Programming on AM is even more limited than that of FM.

Though medium wave is technically capable of long distance reception at night, gross overcrowding of the airwaves in the U.S. makes long distance reception on most frequencies impractical. Long distance reception on the so called "clear" AM frequencies is often poor because the frequencies are no longer actually clear. The program content of the clear medium wave stations is no longer of interest to a wide national audience as in the past. It is an economic fact that local AM and FM stations in low population regions have to serve the largest audiences in their coverage area. The smaller audiences in these areas are not served.

Internet radio suffers from poor quality with slow connections, lack of portability, and the necessity of tying up a computer and Internet connection for radio reception. Big dish satellite radio has excellent quality of reception and no monthly fees are charged. It suffers from lack of portability, complicated tuning that involves moving the dish, and the high cost of the receivers and the dish antenna.

The two satellite services, XM and Sirius offer excellent quality of reception and a one hundred channel programming range that will satisfy most English language listeners. The services cover the entire continental U.S. Receivers cannot be constructed by hobbyists, the cost of receivers is relatively high, and there is a monthly fee charged for each satellite receiver used.

Shortwave radio is the greatest free radio source on earth. Not all broadcasting is in English. But there are numerous English language stations for those interested in other parts of the world or a fresh approach to world news and events. Most stations do not run commercials. Audio quality varies from very poor to good. Tuning shortwave stations on a superhet radio is somewhat more tedious than tuning domestic AM or FM stations. Tuning a regenerative shortwave radio is more involved than tuning a superhet. Modulation on shortwave is mostly mono AM. Though stereo is technically feasible on shortwave, it is rarely used. SSB is used by very few shortwave stations. Audio quality using SSB is acceptable for speech, but not for music reception.

Most shortwave stations broadcast in English for a limited time each day and they change frequencies during a twenty-four hour period. Finding an English language program of a given country or broadcasting service at any given time of day or night is not straightforward. There are no fees for listening; there is a wide range of programming, receivers are inexpensive, portable, and can be built by hobbyists. Thus, in spite of new satellite and Internet broadcasting, shortwave radio remains an attractive service today.

4 Old Regenerative Radio Circuits

The Grid Leak Detector

The circuit of Figure 4-1 is a modern version of the "grid-leak" detector, which was the first application of the vacuum tube. This circuit was discussed in Chapter 1 and is similar to Figure 1-5. In this circuit, a triode vacuum tube replaces the crystal detector of a crystal radio. The grid and cathode of the triode tube function as a vacuum diode which detects the amplitude modulated input signal. The triode provides gain to the input signal in addition to detecting it. The output of the circuit is audio. In many early circuits, headphones were used instead of the plate resistor.

The 1 megohm grid leak resistor in Figure 4-1 was often not used in the original circuits because the effect of the resistor was simulated by leakage through the gas present in the original tubes. When high vacuum tubes became available, the grid leak resistor had to be in place for the circuit to work. In early literature, the resistor was thought of as leaking electrons from the grid of the tube. Today, we explain the circuit by the diode effect of the grid and cathode of the tube and the negative bias produced across the grid resistor by the diode current flowing in one direction through the grid leak resistor.

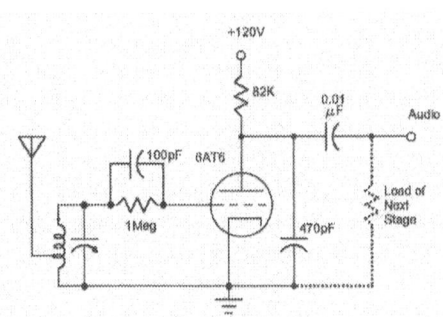

Figure 4-1: A Grid Leak Detector

Figure 4-2A shows the amplitude versus time plot of an amplitude modulated radio signal. For simplicity, the audio modulation shown in this drawing is a single frequency audio tone. Figure 4-2B is the signal appearing at the grid of the triode in Figure 4-1. The audio is represented by the negative carrier peaks. When the signal in Figure 4-2B is low pass filtered by the 470 pF

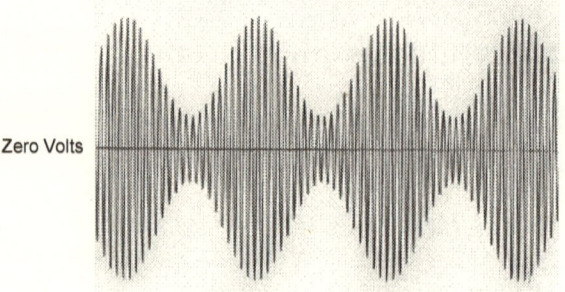

Figure 4-2A: The AM Input Signal to the Grid Leak Detector

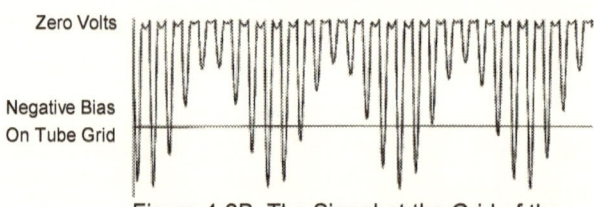

Figure 4-2B: The Signal at the Grid of the Grid Leak Detector

capacitor in the plate circuit of Figure 4-1, the carrier spikes are removed and the result is the audio signal combined with the positive DC plate voltage. If that signal is coupled to the next stage through a coupling capacitor, the DC component is removed and the result is the pure audio signal. This discussion is similar to the discussion of the operation of the diode detector in Figure 1-4.

The average negative DC voltage shown by a line in Figure 4-2B amounts to a reverse bias on the grid of the triode which changes with the carrier level of the received signal. There will be one carrier level at which the triode will generate minimum audio distortion. It is for this reason that the grid leak detector was abandoned in favor of the separate diode circuit in Figure 4-3. In that circuit, the bias on the triode can be kept at its optimum level. In spite of its limitations, this AM detector has been used in most commercial receivers since about 1930.

The signal from the diode circuit in Figure 4-3 passes through the resistor-capacitor π (pi) network that forms a lowpass filter and removes the carrier from the output. The signal then passes through the volume control and a highpass blocking capacitor that removes the DC component from the audio

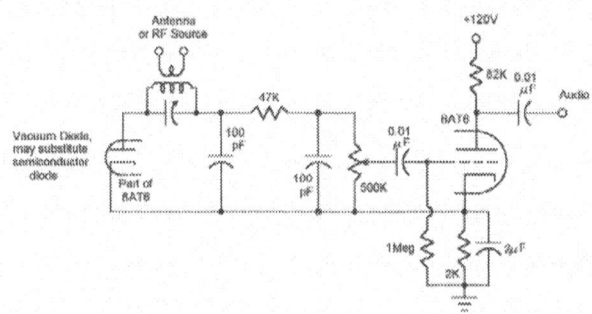

Figure 4-3: A Vacuum Diode AM Detector

signal. The bias on the grid of the triode tube is set by the size of the cathode resistor. This bias remains constant and it is independent of the signal carrier level. In superhets, the negative DC voltage produced by the diode of Figure 4-3 is used for the automatic gain control (AGC) system of the receiver. The diode and triode elements are contained in the same glass envelope of the 6AT6 tube.

The Original Regenerative Radio

Figure 4-4, which is similar to Figure 1-6, shows Edwin Armstrong's modification of the grid-leak detector to form a regenerative radio. A winding in the plate circuit called the "tickler" coil is coupled to the tuning coil in the grid circuit of the tube. This forms an oscillator. The word "tickler" refers to the plate circuit tickling the grid circuit into near oscillation. The feedback of this circuit can be controlled by mechanically rotating the axis of the tickler coil with respect to the tuning coil in the grid circuit. The best operating condition for the receiver is when the angle of the tickler coil is set so that the circuit is almost, but not quite, oscillating. Under this condition, the sensitivity of the detector is very high and the bandwidth of the

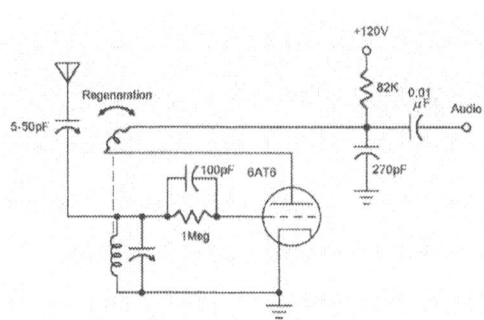

Figure 4-4: The First Regenerative Detector

tuned circuit becomes narrow, much more so than would normally be possible with a single tuned circuit.

When the circuit of Figure 4-4 is adjusted so that it just barely sustains oscillation, it is in its best condition for reception of CW and SSB. For upper sideband SSB, tune the oscillating detector below the signal frequency. For lower sideband SSB, tune the oscillating detector to a frequency above the signal frequency. Careful tuning is necessary to obtain intelligible audio.

For CW, more than one signal will probably be heard simultaneously because the bandwidth of the receiver is not narrow enough to single out one CW station. Each code signal will have a different audio tone and it will be possible to copy one of the signals by concentrating on the audio pitch of that signal. The main tuning control of the receiver can be used to change the pitch of the desired code signal.

The traditional regenerative radio used two variable capacitors in parallel. One of the variable capacitors was large, 100 to 365 pF, allowing the receiver to tune a large portion of the shortwave spectrum with one sweep of that capacitor. The large variable capacitor was called the bandset capacitor and was used to select the shortwave band of interest. Another variable capacitor of about 15 pF (not shown in Figure 4-4) was connected in parallel with the bandset capacitor and was called the bandspread capacitor. The bandspread capacitor was used to tune across a single shortwave band.

Frequency Dials

If the antenna capacitor is removed from the antenna circuit in Figure 4-4, the antenna load on the regenerative circuit will often prevent it from oscillating, regardless of the setting of the regeneration control. The antenna and antenna capacitor are a part of the tuning circuit. The characteristics of the antenna will affect the rotation point of the tuning capacitor at which a given radio station will appear. The setting of the antenna capacitor affects the tuning characteristics of the L-C circuit, as do the characteristics of the antenna.

For that reason, most regenerative receivers do not have a calibrated frequency dial. This is also true of early TRFs and superhets. Early receivers used a "log" dial that showed the numbers 0 to 100 instead of actual frequencies (see

Figure 4-5). Thus, one could "log" the location of a given station on the dial by recording a number from 0 to 100 from the log dial in a "log" book for future reference. The station would remain in the same place on the dial until the antenna or antenna trimmer was changed. When panel space is limited, an alternative to the log dial is a large knob with a numbered skirt, as shown in Figure 7-12. Since the tuning capacitor rotates only 180 degrees, not all of the numbers on the skirt will be used for tuning.

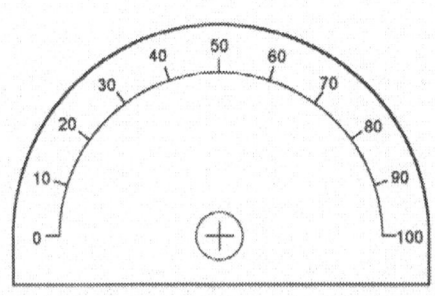

Figure 4-5: "Log" Dial Used for Tuning

Spurious Radiation

As discussed above, the antenna in the traditional regenerative radio was directly connected to the tuned circuit through the antenna trimmer capacitor. When the radio was oscillating, as would be the case when it was tuned for receiving CW, it would transmit a signal over the antenna. When tuning AM stations, the radio would oscillate occasionally while a station was being tuned in.

The spurious radiation of a regenerative radio can be greatly reduced by placing a grounded grid tube buffer between the tuning L-C circuit and the antenna. The buffer tube would provide extra gain to the receiver, but because of the extra cost and circuit complexity, the buffer tube was seldom used.

For military applications, radiation over the receiving antenna meant that the regenerative radio would give away its location. Thus, the radio became unpopular for military applications.

When regenerative radios were used for medium wave broadcast radio, one receiver would interfere with the one next door. This led to the replacement of the regenerative radio by the tuned radio frequency (TRF) receiver for broadcast reception. One example of the TRF was the three tube "neutrodyne" type battery receivers that were popular in the early 1920s. Many different manufacturers made this type of receiver. The radios were used with headphones or a horn type speaker.

DeForest Versus Armstrong

Figure 4-6 shows the DeForest version of the regenerative radio that was used by the U.S. Navy. This is similar to the circuit known today as a Pierce oscillator. This DeForest design was more complicated to tune than the Armstrong version.

Lee DeForest, who invented the vacuum tube, sued Armstrong claiming that the regenerative receiver was DeForest's invention. DeForest won the case, but history gives Armstrong the credit for the invention of the regenerative radio and the circuit revealed in Armstrong's patent has been almost universally used since that time. Accounts of the Armstrong-DeForest controversy often portray Armstrong as a knowledgeable inventor and DeForest as a lucky bungler who happened on the vacuum tube idea. Though DeForest did not completely understand his own vacuum tube device, he did have a PhD degree in physics and his apparently somewhat negative legacy may be due more to his personality than a lack of technical competence.

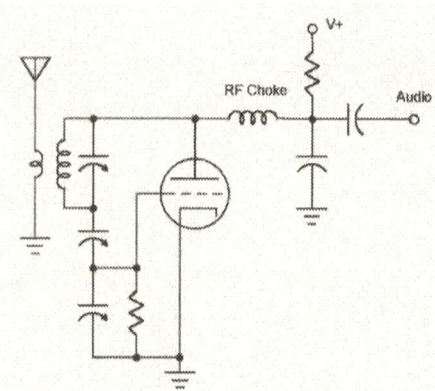

Figure 4-6: DeForest's Regenerative Radio

Passive and Active Regeneration Control

One way to control the regeneration in a regenerative receiver is by varying the portion of the output signal fed back to the input of the circuit. This is a passive control in that it involves the passive (non-amplifying) components of the regenerative circuit. In engineering parlance, the proportion of the output signal fed back is called the "loop gain" of the circuit. This term, loop gain, always has a fractional value and is somewhat of a misnomer. A better term might be feedback attenuation factor. In Figures 4-4, 4-7, and 4-8, the regeneration is controlled by varying the loop gain.

In the active gain control method, the regeneration is varied by controlling the gain of the active device: transistor or tube. In this book, the gain of the active device will be called the "device gain" of the circuit. In Figures 4-9, 4-10, and 4-11, the regeneration is varied by controlling the device gain. In addition to providing a means of regeneration control, the builder must pay attention to both loop gain and device gain when designing a regenerative circuit. Not doing so may lead to some of the regeneration problems discussed below. The product of the loop gain and the device gain must be high enough to permit oscillation of the circuit over the desired frequency range.

The device gain for a regenerative circuit must be kept fairly low and the use of high gain amplifying devices like pentodes and dual gate MOSFETs does not improve the performance of the regenerative circuit. The device gain of the active component has to be reduced to a level that can be easily achieved by a simple triode or JFET circuit. It may be desirable, however, to use pentodes and dual gate MOSFETs. These devices contain an extra control element, screen grid or gate 2, which may be useful in the design.

Problems Associated with the Regeneration Control

- Hysteresis - The radio comes out of oscillation at a lower setting of the regeneration control than that at which the radio went into oscillation. This is sometimes referred to as regeneration control "slop", and it makes tuning the radio considerably more difficult. The cause of hysteresis is, when the circuit goes into oscillation, the large signal generated causes the bias on the tube or transistor to shift. To compensate for the change in bias, the regeneration has to be reduced to a lower setting to bring the circuit out of oscillation.
- Criticalness - A tiny movement of the regeneration control causes the radio to jump into oscillation. This occurs when the device gain of the circuit is too high.

Detuning - Changing the regeneration setting causes the receiver to change the frequency which is received. In the circuit of Figure 4-7, the regeneration is controlled by a variable capacitor which affects received frequency while changing the regeneration level.

Control Noise - This is a scratchy sound heard in the output of the receiver when the regeneration control is varied. The circuit of Figure 4-8 is particularly prone to control noise because the control cannot be bypassed with a capacitor.

Squegging - When the radio breaks into oscillation, a loud squeal is heard in the audio output. There may also be a ringing sound when the radio is set just below oscillation level. The cause is that the radio frequency oscillation turns off and on at an audio frequency. This happens when loop gain of the receiver is too high. Figure 5-9 in the next chapter is prone to this problem. The 7.5 K-ohm resistor in the plate circuit lowers the loop gain to prevent squegging. Lowering the value of the grid resistor can also reduce squegging because that reduces the gain of the circuit at the lower audio frequencies.

Fringe howl - A howling sound is heard when the regeneration control is tuned near the oscillation point. This occurs only in transformer-coupled tube circuits. Figure 5-4 in the next chapter is an example of transformer coupling, but in a transistor circuit. This problem was frequently documented in old radio literature and is caused by the loop gain of the transformer circuit being too high. Fringe howl was often remedied by placing a resistor across the transformer primary to lower the loop gain of the circuit.

Regeneration Control Methods

The first three regeneration control methods covered here are passive methods. The tube retains its maximum gain for amplification of the input signal. The first method of regeneration control, which has already been mentioned, is shown in Figure 4-4. The regeneration is changed by rotating the tickler coil with respect to the tuning coil, thus varying the mutual inductance between the coils. This requires a specially constructed inductor called a vario-coupler. Figure 7-8 shows a drawing of this type of tuning coil. These inductors were once available from radio supply stores. Today, the vario-coupler coil must be home constructed. In early literature, this method of regeneration control was reported to be critical and to cause detuning.

Figure 4-7 shows the "throttling" capacitor method of controlling regeneration. Changing the capacitance changes the amount of radio frequency (RF) bypass to ground at the voltage supply end of the tickler coil. The higher the bypass capacitance, the greater the amount of positive feedback. This method, like

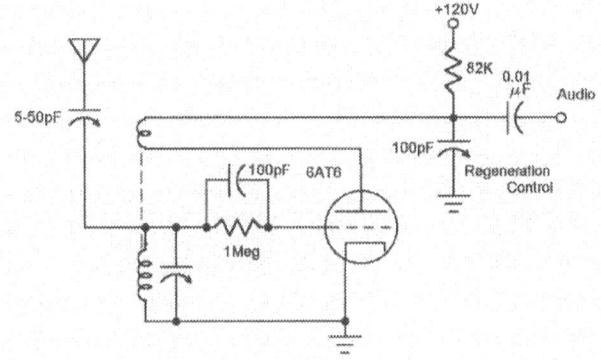

Figure 4-7: Regeneration Control by Throttling Capacitor

the vario-coupler method, had the disadvantage of detuning the radio while the regeneration was being adjusted. The throttling capacitor method provided very smooth and quiet regeneration control, and for that reason it has been popular.

In Figure 4-8, a variable resistor was placed across the tickler coil. This regeneration control does not detune the radio, but it can produce a scratchy noise as the regeneration control is varied. As with the vario-coupler method, the regeneration adjustment may be critical. The variable shunt resistor places an extra load on the tuned circuit, but the effect on selectivity does not appear to be significant.

A method of eliminating the scratchiness is to use a voltage controlled pin diode in place of the potentiometer. A pin diode is a device that changes resistance in accordance with the DC voltage across it. Although a potentiometer is needed to supply the variable DC voltage to the pin diode, the potentiometer can be bypassed with a capacitor to eliminate control noise.

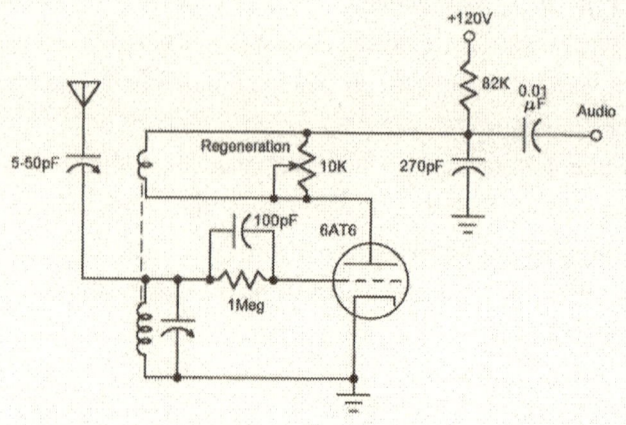

Figure 4-8: Regeneration Control by Variable Resistor

The last three methods of regeneration control are active methods which change the device gain or transconductance of the active device (tube or transistor). Transconductance is a measure of the gain of a tube expressed as the change in output current divided by the change in input voltage. The unit of transconductance is the "mho" (ohm spelled backward) in the U.S. and the "Siemen" in Europe. The mho is a large value of transconductance, so the term micro-mho or µ-mho which is one millionth of a mho, is specified in tube and transistor data sheets. The transconductance of a tube can be altered by changing the operating bias.

This is potentially a good method of regeneration control. It has a negligible affect on the tuning of the radio, it is smooth and quiet, and it is accomplished by varying a voltage. A disadvantage is that when the gain of the tube is reduced to control regeneration, the signal amplification by the tube is also reduced.

In Figure 4-9, the transconductance of a triode is controlled by varying its plate voltage. The plate voltage is changed by a variable resistor in the plate circuit of the triode.

In Figure 4-10, the transconductance of a pentode is changed by changing the screen grid voltage of the tube. An advantage of this circuit is that the current passing through the regeneration control is less than in the circuit of

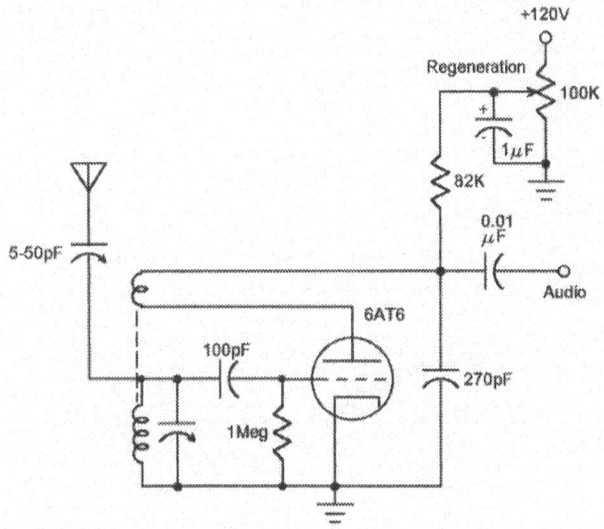

Figure 4-9: Regeneration Control by Changing Plate Voltage

Figure 4-9. The antenna can be attached to the screen grid, thus isolating it from the tuned circuit. Unlike Figure 4-11, the grid leak function is left intact in this circuit and that helps to control the oscillation.

The method of regeneration control shown in Figure 4-11 has been used by the author and has not been encountered elsewhere in literature. Regeneration is controlled by changing the bias on the grid of a triode. Changing the bias level interferes with the detection function of the triode so a semiconductor diode is

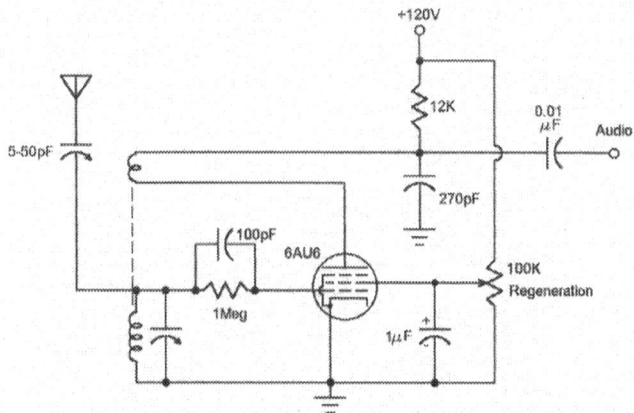

Figure 4-10: Regeneration Control by Changing Screen Voltage

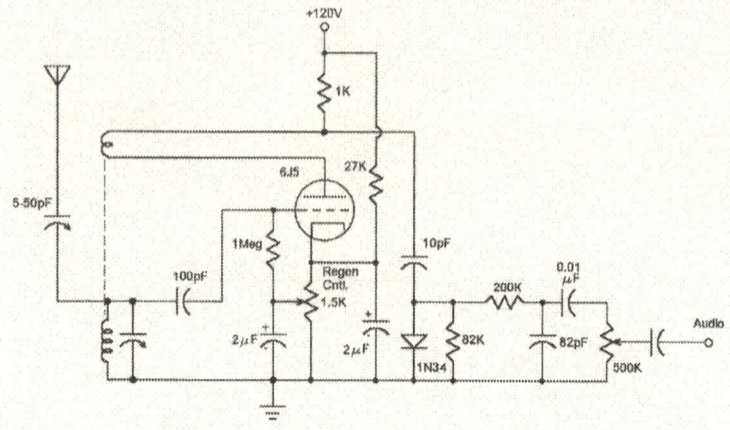

Figure 4-11: Regeneration Control by Changing Control Grid Bias

added to provide AM detection. Alternatively, the RF output at the plate of the tube can go to a wideband amplifier before being detected. As discussed later, a wideband RF amplifier must be constructed with transistors. Grid leak control of oscillations does not function in this circuit. The regeneration control method of Figure 4-11 works well in FET transistor circuits.

5 The New Regenerative Radio

Replacing the Tube in an Old Circuit With a Transistor

The tube in an antique regenerative radio can be replaced, temporarily or permanently with a junction field effect transistor (JFET). This may be desired in order to eliminate the filament power supply, to use a lower supply voltage than is required by a tube, or to replace a defective tube for which a duplicate is not available. The circuit of Figure 5-1 is the usually suggested method of tube replacement. This circuit differs from the original circuit (Figure 4-8) in that a resistor and capacitor are between the source lead of the JFET and ground. The corresponding tube circuit has no cathode resistor. The basic operation of Figure 5-1 is different than the tube circuit it replaces. Computer simulations of this circuit suggest that careful selection of the source resistor is necessary to reduce distortion.

A JFET is not an exact analog of a vacuum tube. The grid and cathode of a cathode grounded tube function both as a diode AM detector and as part of the amplifying triode. The tube is operating in the saturation region of its characteristics. AM detection can also be thought of as occurring because of the bend in the plate current versus grid voltage (transconductance) curve of a tube operating in the saturation region.

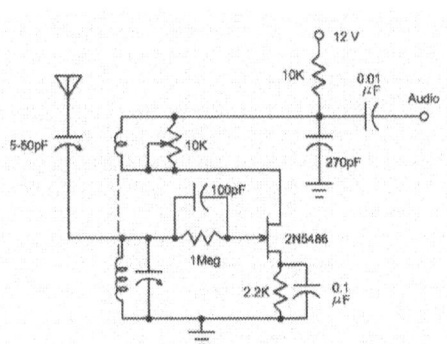

Figure 5-1: Replacing the Vacuum Tube in an Antique Circuit with a JFET

The input diode of a JFET cannot be used as an AM detector because for normal operation of the JFET, the diode must be reverse biased. The bypassed source resistor of the JFET is necessary to bring the device out of hard saturation.

AM detection in Figure 5.1 is accomplished by the bend of the drain current versus gate voltage (transconductance) characteristic of the FET at saturation. There is no diode effect of the FET input as there is with the tube. Ironically, the grid leak resistor and capacitor can be shorted out and the operation of the circuit will not change. The grid leak resistor and capacitor are artifacts from the tube circuit that have no function in the FET circuit.

Figure 5-2 shows a junction FET circuit that works more like the tube version. AM detection is accomplished by an added germanium diode which

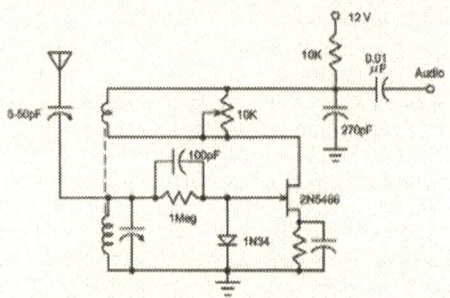

Figure 5-2: An Alternate Method of Replacing a Tube with a JFET

simulates the grid-cathode diode of a tube. The "grid leak" resistor and capacitor function in the same manner as in the tube circuit. The FET is biased to operate in its linear region.

Regenerative Radio Configurations

Versions of the basic tickler coil oscillator circuit shown in Figure 4-4 have been almost exclusively used for regenerative receivers since 1914. This is curious in view of the fact that there are dozens of other designs that will work as regenerative receivers. Almost any RF oscillator in which the positive feedback (regeneration) can be controlled, can be used as a regenerative receiver. Figure 5-3, which is the same as Figure 4-8, will be taken as the starting point of this discussion. The traditional configuration of Figure 4-4 was not used here because that circuit requires a variometer inductor that cannot be readily obtained. Figures 5-3 through 5-7 show five different regenerative radio designs. These configurations can be built with either transistors or vacuum tubes.

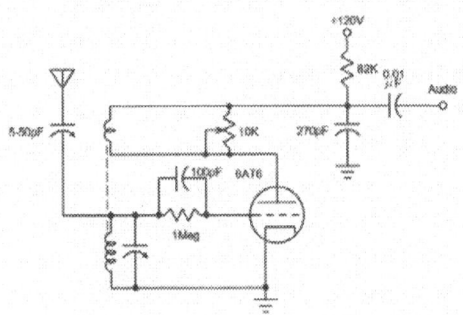

Figure 5-3: Tickler Coil Regenerative Circuit

In Figure 5-4, the L-C tuning circuit is placed in the collector circuit of a bipolar transistor. This is analogous to having the tuning circuit in the plate circuit of a tube. Feedback is accomplished by a small capacitor from the collector to the emitter (plate to cathode of a tube). The regeneration is controlled by the 50 K-ohm potentiometer in the emitter circuit. At least a short (15 foot) indoor antenna and a pair of headphones is required with this radio.

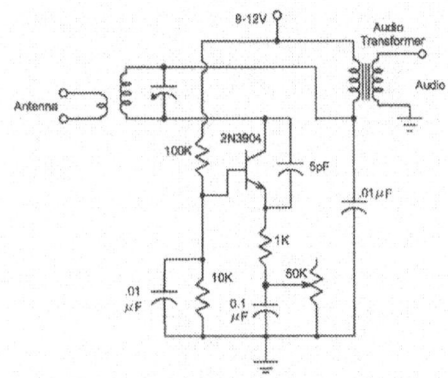

Figure 5-4: Tuned Collector (Tuned Plate) Regenerative Circuit

An outdoor coaxial fed antenna would be better. In order to drive a speaker, extra audio voltage amplification and an audio power amplifier is required.

Figure 5-5 is a basic Pierce oscillator circuit in which the L-C tuning circuit is placed between the plate and grid of the tube. A disadvantage of this circuit is that the tuning capacitor, which must be touched to change received frequencies, is not at ground potential. One remedy for this is to use a non-metallic shaft extension between the capacitor and the tuning knob. Since regeneration in this circuit is controlled by changing grid bias, an extra diode circuit is needed to achieve AM detection.

Figure 5-6 is a Hartley regenerative circuit. Feedback is accomplished

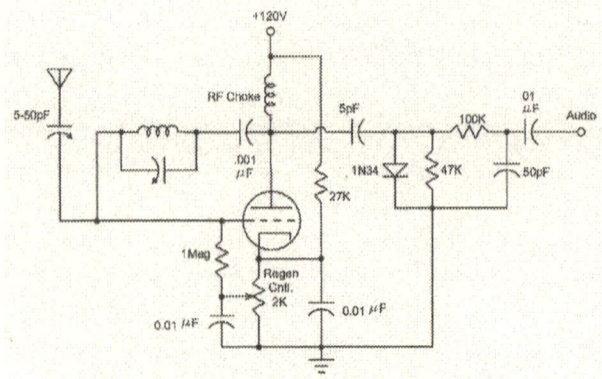

Figure 5-5: Pierce Regenerative Circuit

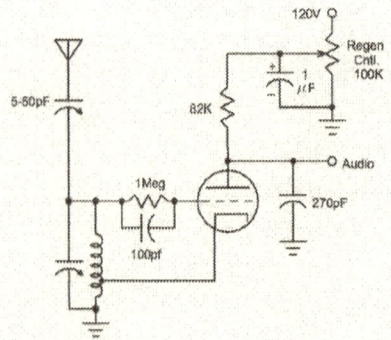

Figure 5-6: Hartley Regenerative Circuit

by connecting the cathode of the tube to a tap near the ground end of the tuning inductor. Regeneration in this circuit is controlled by the variable plate voltage method discussed in Chapter 4. As with the tickler coil circuit, the Hartley circuit requires an especially designed inductor for each shortwave band received. Commercially manufactured inductors cannot be used unless modified. A two pole rotary switch or a plug and socket arrangement is required to select the desired band inductor.

Figure 5-7 is a Colpitts type regenerative receiver. Feedback is from a split capacitor in the tuning circuit to the cathode of the tube. Since grid bias regeneration control is used, a separate diode detector is required.

These are five of many regenerative radio configurations possible. It should be noted that there is more than one possible configuration for the Pierce, Hartley and Colpitts circuits depending on which of the three tube elements the

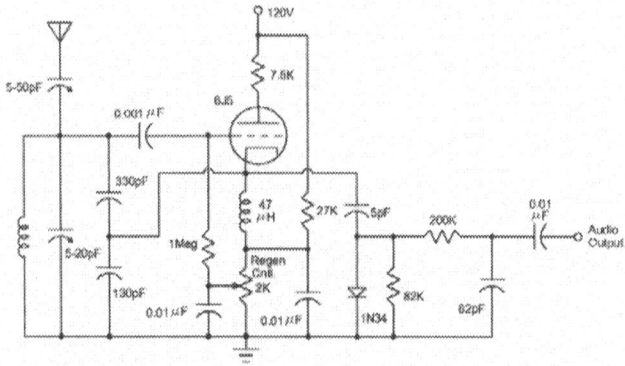

Figure 5-7: Colpitts Regenerative Circuit

feedback is between. When a pentode tube is used, the number of possible radio circuits becomes greater. Using different regeneration control methods expands the number of circuit possibilities further.

Readers who wish to build any of the circuits of Figures 5-3 through 5-7, should note that tube regenerative radios exhibit more variability than transistor circuits. Changing to a different tube of the same type can lead to instability. These circuits should all work, but minor modifications may be needed. The designs are not necessarily optimum.

The tuning inductor in the circuit of 5-7 is a simple two terminal device, one terminal of which is connected to ground. By using this circuit, two terminal inductors for the different shortwave bands can all be attached to ground at one end and the other terminal can be attached to a simple single pole rotary switch or circuit board mounted dip switch. Grounded inductors and simple switching are advantages of this Colpitts circuit.

In the tickler coil design of Figure 5-3, each inductor is a four terminal device and must be individually designed for each shortwave band in order to get the proper amount of feedback between the two coils. When the tickler coil inductor is used, at least three terminals must be switched in order to change bands. Plug in coils are often more convenient in this case.

The formula for determining the inductance of the two terminal inductor required for Figure 5-7 is:

$$L = [1/(2\pi F\sqrt{C})]^2$$

Where:

L = the tuning inductance
π = 3.1416
F = frequency
C = the tuning capacitance

If the circuit capacitance with the variable tuning capacitor at maximum is 150 pF and the lowest frequency to be tuned is 5960 kHz, the required inductance is 4.76 µH. (1 µH = one millionth of a henry). A powdered iron toroidal inductor can be wound that will have a better quality factor (Q) than commercially available solenoids. Table 1 of Chapter 1 lists inductor values for a number of shortwave bands, assuming that the maximum tuning capacitance is about 130 pF and the tuning capacitance range is from approximately 115 pF to 130 pF. Table 2 of Chapter 1 provides details for winding toroidal coils for the various bands.

Block Diagram for All Regenerative Radios

Figure 5-8A is a block diagram applicable to most regenerative radios. In the case of the one tube regenerative radio, the functions of the first four blocks of Figure 5-8A are performed by a single tube. Power amplification (the fifth block) can be accomplished by adding an audio power tube to the circuit.

The five functions of Figure 5-8A can be realized by use of five different circuit modules. The radio design of Figure 11-5 does utilize separate transistors for each of the functions. Separating the different modules allows for the

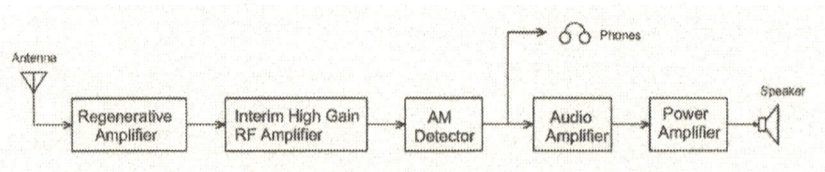

Figure 5-8A: Block Diagram of a Regenerative Receiver

optimization of the individual performance of each function.

Because of the high impedances and large inter-electrode capacitances of vacuum tubes, it is very difficult to design a wide band RF amplifier (3 to 30 mHz) using tubes. Therefore, modular tube regenerative radios, such as the one in Figure 11-10, must have an AM detector directly after the regenerative amplifier The interim amplifier must be a high gain audio frequency (AF) amplifier instead of an RF amplifier. This type of radio is better represented by the block diagram in Figure 5-8B. Can a hybrid radio be built that uses a vacuum tube regenerative amplifier with a transistor RF amplifier? This does work well but it requires a third DC power supply for the transistor circuit .

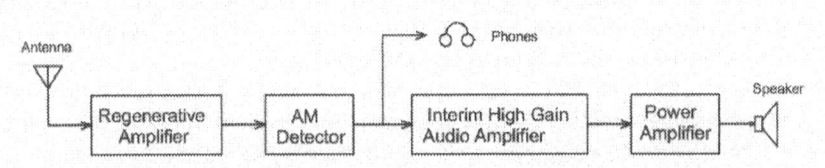

Figure 5-8B: Alternate Block Diagram for Tube Regenerative Receivers

We will now look at some specific circuits representing the different circuit modules in Figures 5-8A and 5-8B.

The Regenerative Amplifier

The output of a regenerative amplifier is a band limited radio frequency (RF) signal instead of an audio frequency (AF) signal. The RF output can be further amplified before the signal is detected. Figure 5-9 shows a Colpitts regenerative amplifier that uses a two terminal bandset inductor. Figure 5-9 is similar to the receiver in Figure 5-7.

The circuit of Figure 5-9 can be built with a dual gate

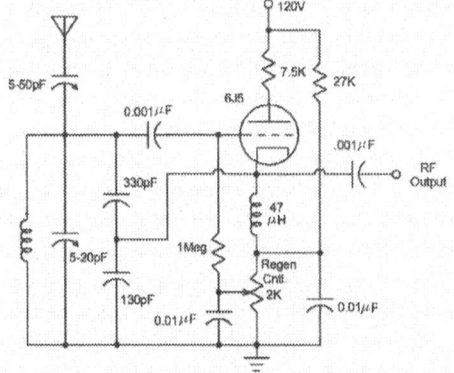

Figure 5-9: A Colpitts Regenerative Amplifier

69

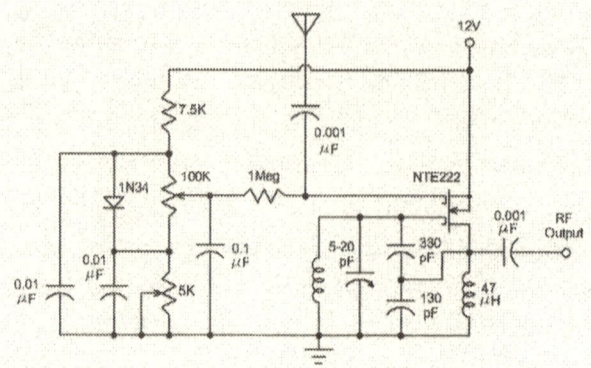

Figure 5-10: A Dual Gate MOSFET Colpitts Regenerative Amplifier

MOSFET as shown in Figure 5-10. This will allow the tuned circuit to be at gate one and the transconductance control and antenna input at gate two. Thus, the antenna is completely isolated from the tuning circuit and the antenna does not very much affect the tuning of the circuit. Furthermore, this arrangement should keep circuit oscillations from reaching the antenna and causing interference with other receivers. This type of isolation can also be accomplished with pentode tube circuits such as the one in Figure 4-10.

Both gates of the dual gate MOSFET in Figure 5-10 have a high input impedance. A short antenna wire can be attached at the antenna input and it will function like an active antenna.

The price of the replacement NTE222 dual gate MOSFETs is currently about ten dollars per unit. This implies that dual gate MOSFETs may be becoming obsolete and unavailable to hobbyists. I have heard of a new type, the BF998, that has a 12 volt breakdown rating. It is apparently a surface mounting type and the price is under $1 each.

The dual JFET version of Figure 5-10, which is shown in Figure 5-11, works well. The two JFETs cost under one dollar each and there is no indication that they are becoming obsolete. The circuit requires a tapped choke in the source lead of the transistor. The same tap can be used for a low impedance (50 ohm) input. Winding instructions for the choke are given in Table 1-2.

Figure 5-12 shows the Hartley vacuum tube version of the regenerative amplifier. Regeneration control in the circuit of Figure 5-12 is smooth, but the simple, two terminal inductors of the Colpitts circuit in Figure 5-9 must be

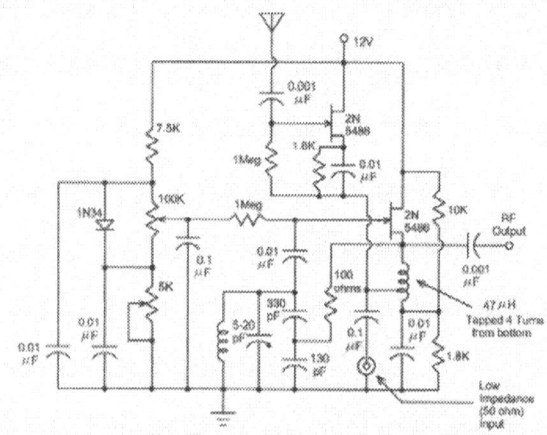

Figure 5-11: JFET Version of Colpitts Regenerative Amplifier

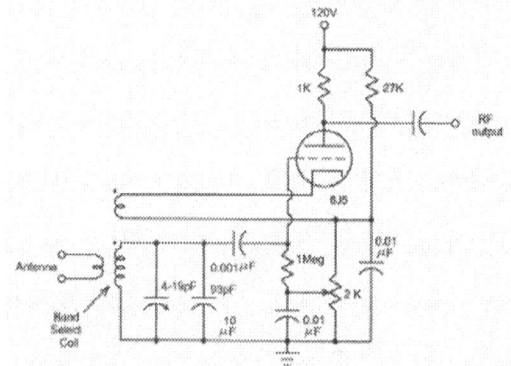

Figure 5-12: A Hartley Regenerative Amplifier

sacrificed. Changing shortwave bands of the circuit of Figure 5-12 will be more complicated. Indeed, the band inductor shown in Figure 5-12 is a six terminal device.

One problem with Colpitts and Hartley type regenerative circuits is getting a significant signal out of the circuit without loading the tuned circuit. Tuned circuit loading reduces the selectivity of the radio. In Figure 5-13, a cathode follower is used to couple the large signal at the grid of the input triode to the output without appreciably loading the grid circuit of the regenerative amplifier. This requires an additional triode, but a tube containing two triodes in one envelope like the 6SN7 can be used. The output of Figure 5-13 is low impedance (700 ohms) and is not very sensitive to loading by the next stage.

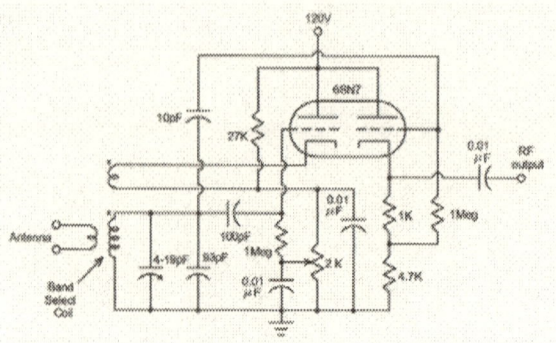

Figure 5-13: A Hartley Regenerative Amplifier with Buffer

The Interim Amplifier

Since the Colpitts regenerative amplifiers of Figures 5-9 through 5-11 provide less output voltage than the equivalent tickler coil circuit, RF amplification following the regenerative amplifier is desirable. Several IC circuits with variable gain have been investigated for this application. These circuits were all unstable and would oscillate at some gain settings. The simple four transistor circuit of Figure 5-14 works well. Each of the four stages provides 20 DB of gain. The last stage has a switch which allows a change in gain of that stage from 0 DB to 20 DB. Thus, the overall gain of the amplifier can be either 60 DB, or 80

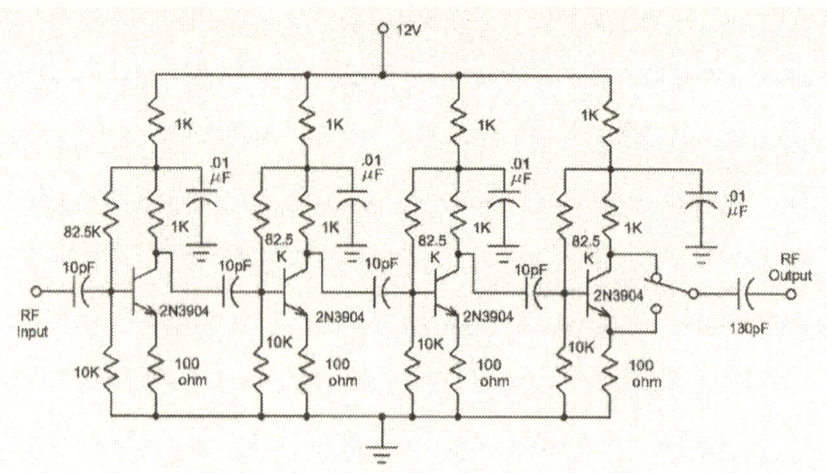

Figure 5-14: An Interim Wide Band RF Amplifier

DB. Twenty DB is equal to a gain of 10, 60 DB is a gain of 1000, and 80 DB is a gain of 10,000. The most commonly used gain setting of the circuit in Figure 5-14 is 60 DB.

As mentioned in the section on the regenerative radio block diagram, it is not practical to build a wide band RF amplifier using vacuum tubes. Therefore, in tube regenerative radios, the regenerative amplifier must be followed by an AM detector and the interim amplifier must be an audio amplifier, preferably with a gain of 1000 (60 db) or more. The block diagram of Figure 5-8B is applicable.

Figure 5-15 shows a high gain audio amplifier using a pentode tube with a transistor active plate load. The transistor does not provide any amplification to the circuit. Instead the transistor provides a high active load impedance to the pentode plate circuit while supplying the 3.2 milliamperes of DC plate current required by the tube.

The pentode needs a high plate impedance in order to produce a high gain. The voltage at the base of the transistor must be correct to within a few hundredths of a volt and the bias may drift over an hour or so. This circuit works well for homebuilt radios, but would not be practical for commercial equipment, due to the necessity of occasionally readjusting the bias with the 500 ohm trimmer resistor. The gain of the circuit is about 250, or 48 DB.

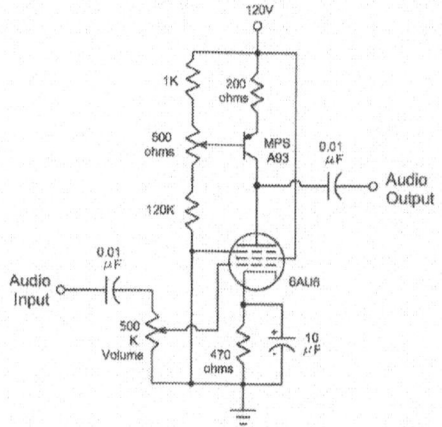

Figure 5-15: A High Gain Interim Audio Amplifier

Some readers may object to having a transistor in a vacuum tube circuit. The circuit of Figure 5-16 uses a dual triode tube. In this circuit, the bias does not drift and no adjustments are required. The total gain is about 3200 or 70 DB. Since both triodes are in the same glass envelope, the pair looks about the same as a single tube.

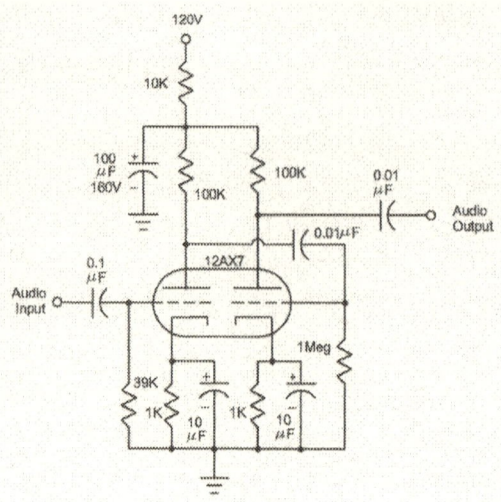

Figure 5-16: Dual Triode Interim Audio Amplifier

The AM Detector

The triode "grid leak" AM detector has been previously discussed (Figures 1-5 and 4-1) and will not be covered here. Also, as previously mentioned, there are other AM detectors such as the envelope detector and the precision rectifier, which demodulate AM with lower distortion than the diode circuit. These are useful when high quality AM reception is desired. The quality of received signals on the shortwave bands usually does not warrant the use of low distortion AM detectors. High sound quality broadcasting and stereo are possible on the shortwave frequencies, but it is not currently provided by broadcasters.

The diode AM detector has been used almost exclusively for commercial circuits for the last eighty years. The diode detector, which is most often used in tube superheterodyne receivers, is shown in Figure 5-17. For this version, the voltage level at the cathode of the diode must be at the same DC level as the output ground. Otherwise, the diode will not function as a switch as desired. The intermediate frequency (IF) input to this detector is usually from the IF transformer of the final IF amplifier stage of the superhet. The detector loads the

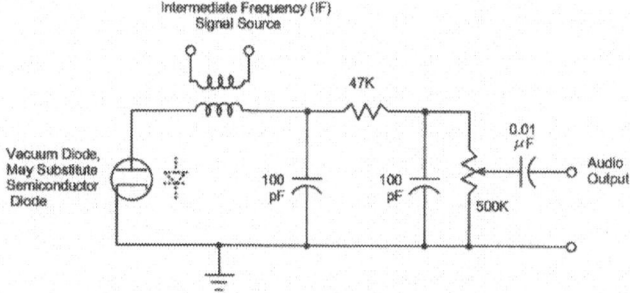

Figure 5-17: Detector Used in Most Commercial AM Receivers

output of the final IF amplifier somewhat. The input impedance of the following audio stage should be 500 K-ohms or more.

Like the detector circuit in Figure 5-17, the semiconductor diode circuit of Figure 5-18A requires that the input RF signal be at the same DC voltage level as the ground terminal of the detector. This condition is met when the input to the circuit is taken from an inductor or transformer. The circuit of Figure 5-18A is often used with tube type regenerative receivers that have been modified so that

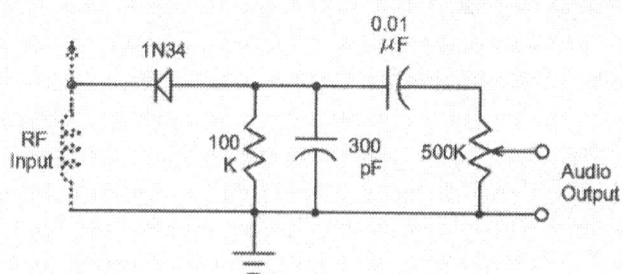

Figure 5-18A: Detector for Use When Source is at Zero DC Offset Voltage

the triode no longer functions as a detector. Since the semiconductor diode does not require filament power, the complexity of the circuit is not greatly increased.

In the detector of Figure 5-18B, the DC voltage level of the AC source does not have to be the same as the return of the output of the detector. The voltage across the input capacitor adjusts to provide the proper switching level for the diode. This is convenient when the RF signal is coming from the plate of another tube.

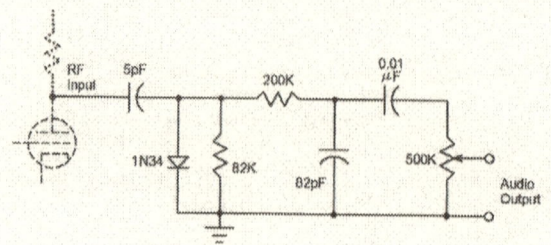

Figure 5-18B: AM Detector That Works When There is a DC Component in the Input RF Signal

The detector of Figure 5-19 was developed by the author for use in transistor receivers. As with Figure 5-18A, the RF source in Figure 5-19 does not have to be at zero DC level. The input impedance of this circuit is high and it does not appreciably load the RF source. The output impedance of Figure 5-19 is low and the audio load of the next stage or headphones can be as small as 4 K-ohms. Since the diode is operated with a small forward bias, a silicon diode such as the 1N914 will work as well for this application as a germanium or hot carrier diode.

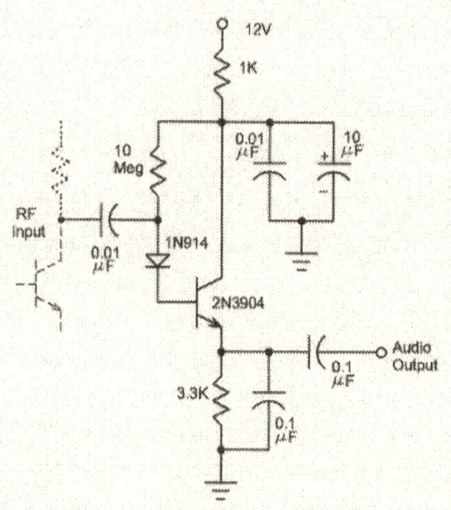

Figure 5-19: Non Loading Detector That Works With an Input DC Offset

With these differences noted, the circuit of Figure 5-19 operates like other diode detectors. Analysis of this circuit might suggest that the circuit would function the same if the diode was shorted out. However, the diode is necessary.

The Audio Power Amplifier

A volume control not shown in Figure 5-8 usually follows the AM detector. The volume control can be a conventional potentiometer or an electronic attenuator. The detector of Figure 5-19 can directly drive the power amplifier of Figure 5-20 without the extra audio amplification shown in Figure 5-8A.

The LM386 IC in Figure 5-20 is over thirty years old and has been used on U.S. spacecraft. In spite of its age, the part is widely available in several different versions. Some later audio amp ICs produce technically better sound

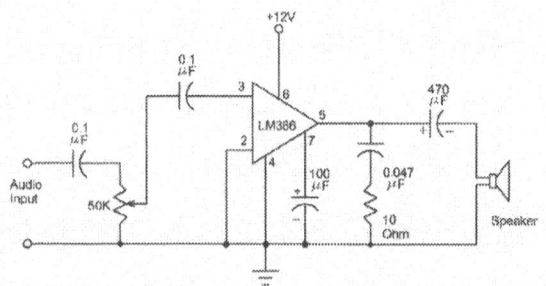

Figure 5-20: An Integrated Circuit Audio Power Amplifier

quality, but the LM386 class B amplifier sounds very good, it is less complicated to use, it will operate with a single power supply, and it uses a convenient supply voltage.

The preferred speaker is at least 4 inches in size and has an impedance of 8 ohms. The 300 milliwatt output of the LM386 is adequate for most small radio applications. If more power is desired, a one watt power amplifier, the LM386A, is available. Figure 5-20 shows the LM386 power amplifier set for a voltage gain of 20.

The single-ended tube power amplifier in Figure 5-21 uses a 6AQ5 or 6V6 tube with a plate voltage of 120 volts and can provide about 450 milliwatts of output power. The output power can be increased to 3 watts if the plate voltage is increased to over 300 volts.

A tube such as a 50C5 or 35C5 can produce an output of

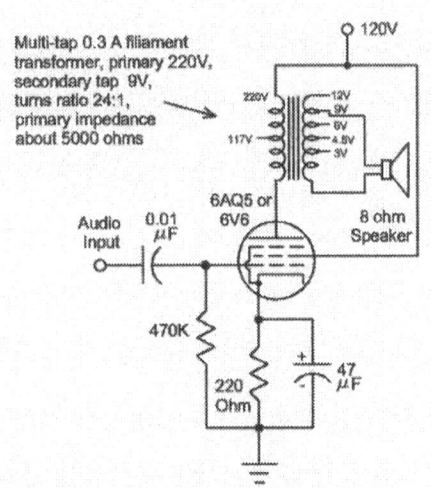

Figure 5-21: A Vacuum Tube Power Amplifier

more than one watt using a 120 volt plate supply. However, the filament voltages (50 volts and 35 volts) required for these tubes cannot be conveniently obtained from filament transformers. The distortion level for these tube circuits is considerably higher than for the IC circuit of Figure 5-20. However, when used for shortwave radio reception, the difference in sound quality is not readily apparent.

6 Other Circuits Using the Regenerative Principle

RF Oscillators

To receive code and SSB, the regeneration control on a regenerative receiver is turned up until the circuit oscillates continuously, at a frequency slightly different from the non-oscillating receiver frequency. A regenerative radio can be used as a variable frequency RF oscillator, as shown in Figure 6-1. The circuit is called a "tickler coil" oscillator, a name that survives from the early radio era. Another name for the circuit is a "tuned grid oscillator". Other configurations of the regenerative radio such as the Colpitts and Hartley circuits can likewise be used as RF oscillators.

This can be useful when a laboratory RF generator is not available to the radio experimenter. Any regenerative receiver can be used as an RF oscillator, or a one transistor oscillator can be easily constructed. If an inexpensive frequency counter such as the one shown in Figure 10-7 is available to the experimenter, this simple RF generator can be tuned with great precision.

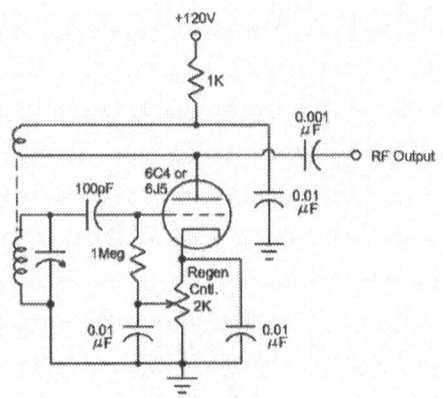

Figure 6-1: A Tickler Coil RF Oscillator

Rudimentary AM Transmitter

In a similar application of the regenerative radio circuit, the RF oscillator described above is AM modulated by a separate tube to form a miniature AM radio station. The regeneration control is turned up until the circuit oscillates continuously at a frequency within the medium wave AM band. The circuit is shown in Figure 6-2. The primary of a speaker transformer can be

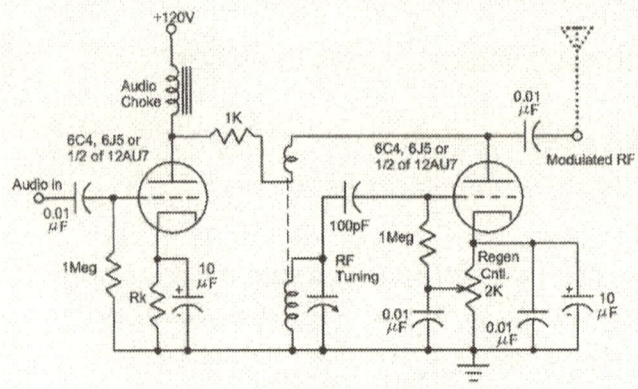

Figure 6-2: A Low Power AM Transmitter

used as the audio choke. The output can be coupled to an antenna and picked up on any AM radio.

This circuit has been referred to in past literature as a "phono oscillator". This name assumes that the audio source is a phonograph pickup. The RF power generated is only a few tens of milliwatts and the FCC allows the use of that much power without having a transmitting license. Broadcasting with this circuit is legal on the medium wave band, but would not be legal on shortwave. To operate at power levels greater than the legal limit, the circuit would have to be followed by a linear power RF amplifier.

The Q-Multiplier

When the regeneration control of a regenerative radio is set just below the oscillation point, the setting at which a regenerative receiver is normally operated for AM reception, the circuit can be used as a high gain, high selectivity one stage RF amplifier (see the section on regenerative amplifiers in Chapter 5). For this purpose, the operating bias on the RF amplifier tube or transistor must be selected so that the circuit stays within its linear range and does not detect the audio signal.

This application of the regenerative principle is called a Q-multiplier and is shown in Figure 6-3. A regenerative oscillator circuit (bottom tube) is added to the RF amplifier stage (upper tube) of an existing superhet receiver. The combination of these two circuits forms a regenerative RF amplifier stage that has a much narrower bandwidth than the original RF amplifier. Thus, the Q-multiplier provides extra gain and selectivity to the input stage of the superhet.

The regeneration control is used to select the desired bandwidth and the control must be set so that the RF amplifier stage is below the oscillation point. A small variable capacitor (not shown) connected in parallel with the tuning capacitor of Figure 6-3 can be used to tune the input stage within the IF bandwidth of the superhet receiver.

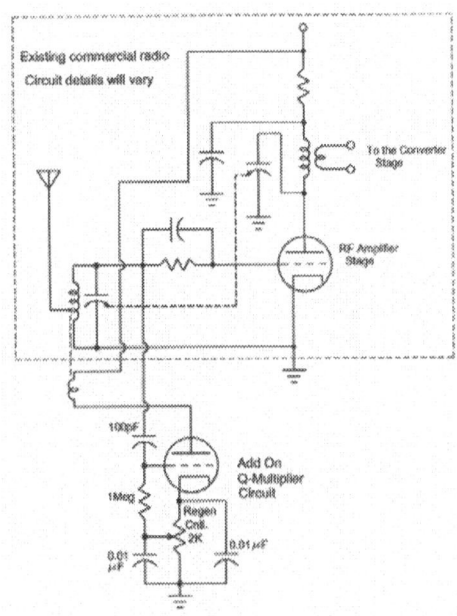

Figure 6-3: A Q-Multiplier Add On Circuit

The Regenerative IF Amplifier

Another application of the regenerative RF amplifier is a variable bandwidth IF amplifier in superhet receivers. The circuit in Figure 6-4 shows how the regenerative IF amplifier might be constructed. Like the circuit in Figure 6-3, this amplifier has been referred to as a Q-multiplier in some literature.

Since the IF frequency (often 455 kHz) does not change, the setting of the regeneration control will not need to be adjusted except when it is desired to alter the bandwidth. This circuit can be used to achieve an IF bandwidth of only a few hundred cycles, which is desirable when CW code is being received.

This circuit could also provide a BFO function if the regeneration control was set for continuous oscillation. However, a separate dedicated BFO

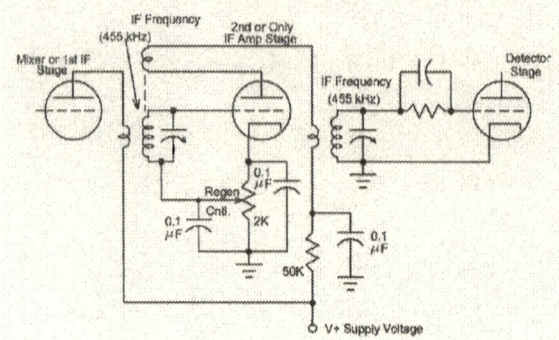

Figure 6-4: A Regenerative IF Stage in a
Superheterodyne Receiver

stage would allow for separate adjustment of bandwidth and BFO.

7 Mechanical Aspects of Radio Design

The Breadboard Type of Radio Construction

The first tube circuits were built on a wooden board which was, and still is referred to as a breadboard. Tube sockets were the surface mounting type (now rare) that were fastened to the board with wood screws. Tie points were attached to the board to secure resistors and capacitors. Open frame transformers could easily be attached to the wood surface. A vertical plastic panel was often attached to the front edge of the breadboard and was used to mount variable capacitors controls and jacks. The wiring between components was above the upper surface of the breadboard. There was usually no wiring beneath the board.

Early commercial TRF receivers were constructed in this way. Sometimes the circuit was on a sheet of phenolic plastic which was mounted over a wooden base. Often, the plastic circuit board was placed at the bottom of a wooden box with a hinged top cover. The front panel was the front surface of the box. The plastic circuit board was similar to the modern printed circuits except that conductors were sheet metal strips riveted to the board instead of traces bonded directly to the plastic.

Chassis Type of Construction

Toward the end of the 1920s the sheet metal chassis replaced the breadboard for circuit construction. The chassis was a rectangular metal box which was used with the open side down. Tubes, transformers, and other components were mounted in holes cut into the top of the chassis. Controls were mounted horizontally on the front side of the chassis. Figure 7-1 is an example of chassis construction (The circuit is an audio pre-amplifier) . A front panel, if used, was attached to the front surface of the chassis. The panel was held in place by the mounting nuts of control bushings passing through both the front chassis and the panel. A cheap chassis can be made with an inverted metal cake pan, but the appearance of this is not good. Often the main reason for building vacuum

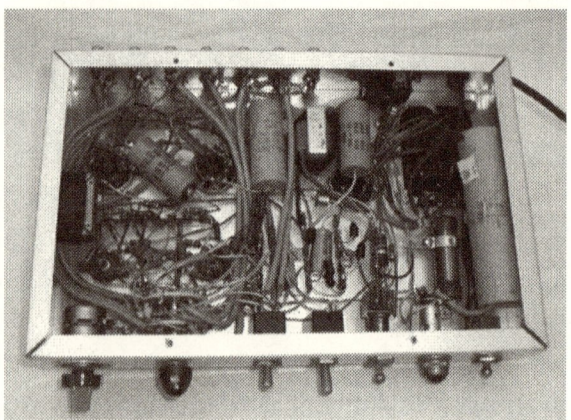

Figure 7-1: An Example of Metal Chassis
Construction

tube circuits is the exotic appearance of the equipment. Aluminum chassis can still be purchased from Mouser Electronics and other suppliers. The one shown in Figure 7-1 is a Bud Metals model AC-407 and it currently sells for about fifteen dollars. With access to a sheet metal shop, a similar chassis could be built cheaply using scrap sheet metal.

Holes are drilled in the sheet metal for each component mounted. Chassis mounted tube sockets and insulated tie points are attached to the metal chassis using machine screws and nuts. Small holes are made in the chassis with an ordinary electric drill.

The mounting of tube sockets requires round holes that are too large to be made with a drill. One way the large holes are made is with a punch and die device called a chassis punch. These tools are available in various shapes and sizes. Circular punches are the most common. Chassis punches are still available, but they are expensive (see Figure 7-2).

A slower but cheaper way to make a large hole in chassis sheet metal is to drill several smaller holes inside the outline for the large hole; then use a circular file to enlarge the hole to the proper size and shape.

Another tool, called a nibbler, cuts one small bit from the sheet metal at a time. A circular hole is drilled inside the area to be cut out. The nibbler is inserted in the hole and the desired outline is cut one "nibble" at a time. The nibbler works better for square holes than round ones and it is nice for cutting holes for chassis type transformers. The tool is fairly inexpensive and is widely available. Figure 7-2 shows the nibbler tool on the bottom left.

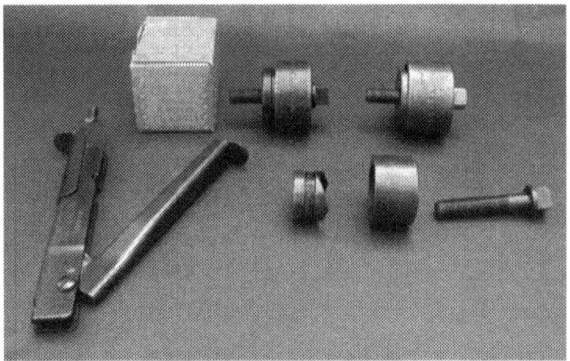

Figure 7-2: A Nibbling Tool and Several Chassis Punches

The chassis is the preferred base for tube equipment. The metal chassis is used as a common electrical ground point for the entire circuit. Furthermore, the metal chassis provides electrical shielding for the circuit.

Printed Circuits

On a printed circuit board, the circuit conductors are bonded directly to the plastic surface of the board. The single sided circuit board was most common during the first two decades that boards were used. The board usually starts as a plastic sheet completely covered with a layer of copper conductor on one side. The circuit pattern is produced by masking the part of the copper that is to remain after etching. The unmasked copper is etched away in a chemical bath. The masking is removed with steel wool. Holes are drilled through component pads and components are mounted by passing the leads through the holes on the plain plastic side of the board and soldering them to the metal trace on the opposite side of the board.

Printed circuit boards became popular in the 1960s and a board from this era is shown in Figure 7-3. These boards were not a good choice for use with vacuum tubes. The heat from the tubes could damage the plastic, causing the

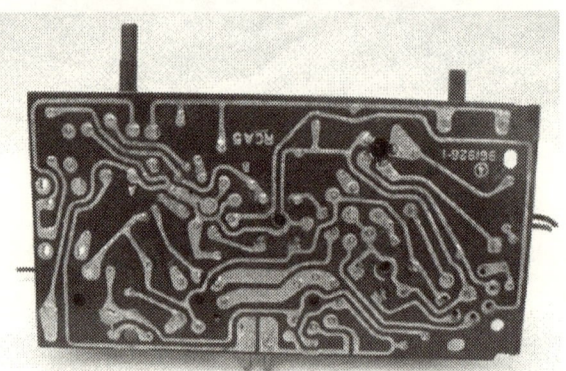

Figure 7-3: A Vacuum Tube Radio on a Printed Circuit Board

metal traces to separate. Also, the trauma of plugging and unplugging the tubes into sockets mounted on the plastic board tended to crack the board. Heavy parts like transformers mounted on printed circuit boards would crack the board if the device was dropped. Most surviving tube printed circuit boards from the 1960s are cracked, including the one shown in Figure 7-3. Nevertheless, the cost of the printed board was considerably less than for chassis construction. Printed circuits were used for tube circuits in radios and televisions until the end of the tube era.

Printed circuits brought vacuum tube circuits in a full circle back to a breadboard like design. In order to provide mechanical support for the printed circuit board, the edges of the board were inserted into slots provided in a plastic cabinet of the radio or TV. The shielding provided by a metal chassis was no longer present.

Printed circuit boards are well suited for transistor and IC circuits and printed circuits are used universally today for manufactured solid state circuits. A PC circuit designed by the author is shown in Figures 7-4A and 7-4B. The type of PC board so far described is the "through the hole" type board, and it requires

Figure 7-4A: A Transistor and Integrated Circuit Printed Circuit Board

manual assembly during manufacturing. The double sided PC board has traces on both sides and allows for more circuit complexity. Commercial printed circuits often have two or more layers. Multi-layer boards are created by laminating sub-boards together. The internal traces are accessed by plated-through holes called vias.

Single sided printed circuit boards are usually adequate for hobbyist circuits. Hobbyist printed circuit boards can be made at home. The simplest method of creating the resist pattern is to apply pads and traces directly to the copper board from wax transfer sheets purchased for the purpose. The board is then etched and the wax transfers are scrubbed off. A disadvantage of this method is that it creates only a single board and computer generated patterns (as in Figure 11-7) cannot be used directly.

There are many computer software programs that can be used to create the resist pattern. Several of these programs are available free on the Internet. The software that I use is a proprietary program called "PC Boards". Figure 7-4B shows an etched PC board created with this program.

Figure 7-4B: Underside of the Printed Circuit Board Before the Parts are Inserted

The easiest method of putting a computer generated resist pattern on a copper clad board is the toner transfer method. A laser printer prints a mirror image of the pattern on a special plastic transfer sheet. The sheet is placed on the foil side of a copper plated board and heat from an electric clothes iron is used to transfer the toner from the transfer sheet to the copper side of the board. The parts of the copper not protected by toner resist are etched away with a ferric chloride solution. Holes are drilled in the board at the center of the component pads. The computer generated pattern may be used to create as many identical boards as desired. Printed boards are time consuming to develop and construct. However, printed boards allow for compact and neat placement of parts and are worth the effort for some home projects.

In a newer manufacturing process called surface mount technology (SMT), there are no holes in the circuit board and special SMT electronic components are soldered to the foil side of the board. Some method must be used to hold the components to the board until the solder is applied. SMT is suited to automated manufacturing, but is difficult for hobbyists. Construction of SMT printed boards subjects the electronic parts to higher temperatures than the "through the hole" method. If a needed part is only available in SMT configuration, the part can be soldered to the foil side of a "through the hole" printed board. Manually attaching a single SMT part to a board is not hard, but holding a whole board full of SMT components in place while they are soldered is difficult.

Prototyping Tube Circuits

Vacuum tube prototype boards are rare today. A tube prototyping board must be used in conjunction with a power supply having outputs for both tube filament and the tube plate supply voltage. I do not have a tube prototyping setup.

A breadboard setup (attaching all components to a wooden board with wood screws) might be a reasonable means for prototyping tube circuits. I develop new vacuum tube circuits by soldering and unsoldering connections under a conventional chassis. When the final circuit configuration is determined, the chassis circuit is cleaned up and resoldered. This is cumbersome, inconvenient, and may require changes to the metal chassis. An advantage of this method, is that when the design is finalized, the tube project is close to completion.

Semiconductor Prototyping

The dearth of prototyping methods for vacuum tube circuits is countered by a plethora of methods for transistor and integrated circuits. I use a perforated board called vector board. Springy metal clips called vector clips are inserted into the holes of the board to hold the components in place. The clips are intended to be wired and soldered below the board. I do not do much wiring on the underside of the vector board because I like to be able to see the whole circuit from one side. The top of the vector clips are supposed to hold components without soldering; however, it is a good practice to solder components to the top of the clips. A completed vector board circuit is as permanent as a PC board, though not as neat. It can be used as the completed project. An example of a vector board circuit is shown in Figure 7-5.

One type of prototype printed circuit board that requires soldering has holes properly spaced for standard IC pins. The solder pads under the board have to be connected with wire. There are many prototype boards that do not require soldering. Connections are made by inserting wires into socket-like holes in the board. Nearly every supplier of electronic parts, including Radio Shack, has some prototyping board to offer.

Figure 7-5: A Transistor and IC Circuit Fabricated on Vector Board

The only non-soldering prototyping method that I have found to be reliable is a method that utilizes springs embedded in cardboard (Figure 7-6). This is often seen on educational electronics board kits for children. It is not easy

Figure 7-6: Circuit Connections on a Spring Terminal Board

to get wires meshed into the springs. A pair of needle nose pliers is helpful. But once the spring connection is made it is almost as permanent as a soldered connection.

Variable Capacitors

Figure 7-7 shows several variable capacitors. While a myriad of air variable tuning capacitors are available from surplus sources, finding a particular

Figure 7-7: Variable Tuning Capacitors
(The Second Capacitor From the Left and the One
at Bottom Right Have Built In Reduction Drives)

size at any one time is difficult. The variable tuning capacitor of a particular size and characteristic needed for shortwave receivers is the most difficult item to obtain in today's market. The most commonly available variable capacitors are the 365 pF type that were used in medium wave broadcast band circuits. These capacitors are too large for most shortwave applications. Also, the capacitance versus rotation characteristic is usually not linear.

Many transistor radios of the 1960s and later, contain plastic insulated variable capacitors that are smaller than the air variable counterparts. These variable capacitors are not as robust as air capacitors and they are prone to intermittent and imprecise tuning. They are usually in the 150 to 365 pF range, but those from AM/FM radios will have a smaller capacitance FM section. A major disadvantage of plastic capacitors is that they are not made for chassis mounting and they do not have standard 1/4 inch shafts. They must either be used with tiny tuning knobs or adapted to a standard shaft size.

The variable tuning capacitor I use most often for shortwave tuning is the 5 pF to 20 pF size. Some types of variable capacitors can be reduced to a lower capacitance by pulling out some of the movable plates, but I do not like doing that. A variable capacitor can also be reduced in capacitance value by adding a fixed capacitor in series with it. The problem with the series fixed capacitor is that the frequency versus rotation characteristic of the combination of fixed and variable capacitors will be nonlinear. This technique, however, can be used where only a small decrease in maximum capacitance is required.

There are new air variable capacitors being manufactured for microwave ovens and such, but these capacitors have not been made available to hobbyists. For those reasons, the reader may consider building his own variable tuning capacitors. A good source of information on this is the Internet where plans for homemade capacitors can be found. Check the site of Lindsay Publications. The unavailability of suitable new variable capacitors for homebuilt receivers might present a small business opportunity for interested readers. The reader could build variable capacitors for hobbyists.

An alternative method of variable capacitive tuning that will be introduced later is the use of voltage controlled varactor diodes.

Mechanical Reduction Drives

A mechanical reduction drive, with or without an attached dial, can be connected to the tuning capacitor of a shortwave radio. Several full turns of the reduction drive knob produces a 180 degree sweep of the capacitor shaft. A drive ratio of three full turns to one 180 degree sweep (called a 3:1 ratio) is good for most shortwave applications. Sometimes, a reduction drive is built right into a variable capacitor. In Figure 7-7, the second capacitor from the left, and the capacitor in the lower right have built in reduction drives. Reduction drives were once manufactured in large numbers and were available at reasonable prices. They made tuning crowded sections of the shortwave bands easier. The drives vary in quality. One failing of drives is called backlash. When changing tuning directions, there is a dead zone of a degree or so in which turning the shaft produces no movement of the variable capacitor. More expensive drives eliminate this flaw.

Reduction drives are rare, but can still be purchased from surplus sources at prices of ten dollars and up. New precision drives from England cost over thirty dollars. Other methods of producing slow tuning are by use of a dial cord and pulleys, gears, and friction drive arrangements. The Internet (the Lindsay site and others) has information on home building mechanical variable capacitor drive devices.

In the circuits of this book, fairly slow tuning rates are obtained by electrical means. A small variable tuning capacitor is placed in parallel with a larger fixed capacitor. Changing from one shortwave band to another is accomplished by switching a different inductor into the circuit. The shortwave bands, which are 500 kHz to 750 kHz in width, are each covered by one sweep of the tuning capacitor. On higher frequency bands, the radio will tune considerably outside the frequency limits of the international band.

When varactor diodes are used for capacitance tuning, the control device is a potentiometer which has a rotation angle of about 300 degrees as compared to the 180 degrees of a variable capacitor. Using the potentiometer with its wider sweep has an effect equivalent to mechanical reduction.

Variometers and Other Inductors

An inductor that has completely disappeared from the market is the variometer (Figure 7-8). This is two coils, one of which rotates or slides inside the other. There are two types of variometers: the variable tuning inductor type and the tickler coil type.

In the variable inductor type of variometer, the two coils have approximately the same inductance and the windings are connected in series. When the coils are aligned so that the turns are in the same direction, the total inductance is maximum. When the inside coil is rotated 180 degrees, the windings of the two coils are in opposite directions and the total inductance is minimum. Thus, the variometer provides a variable inductance that can be used instead of a variable capacitor for tuning. The inductance is changed by rotating the inside coil with respect to the fixed outside coil. The Q factor of the coil will be lowest at the lowest inductance setting.

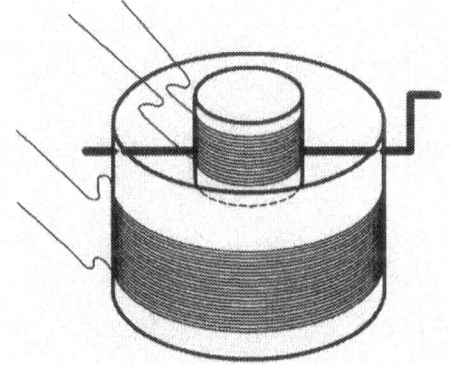

Figure 7-8: A Variometer Inductor

These variable inductors were used with fixed capacitance for tuning in early radios. They were tuned to different stations by changing the inductance instead of the capacitance. In commercial versions of the variometer, the two coils were wound on spherical coil forms with the inside sphere rotating within the outside sphere. Thus, coupling between the coils was maximized. Concentric spherical coil forms are not easy to construct and it is difficult to wind wire upon a spherical surface.

The tickler coil variometer was used only in the regenerative radio. The inside coil of the variometer had fewer turns than the outside coil and this coil was used as the tickler coil of the radio. Close coupling between the two coils is less important than for the variable inductor variometer. Regeneration is controlled by rotating the tickler coil through an angle of 90 degrees to vary the coupling to the main tuning coil from minimum to maximum. Intuitively, this would seem to be a good method of controlling regeneration. However, early radio articles state that this method results in detuning the receiver as the regeneration is changed. Also, this regeneration control is said to be "critical", that is, the transition from non-oscillating to oscillating is not smooth.

Since manufactured variometers are currently unavailable, the reader might want to build his own. Details are available at various sites on the Internet. Building variometers for the hobbyist market might be another small business opportunity for readers.

Several types of inductors are shown in Figure 7-9. The coils used in most old regenerative radios were air core plug in coils wound on a one-inch

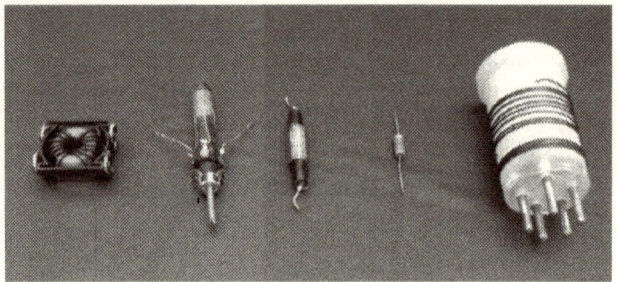

Figure 7-9: Various Inductors

cylindrical coil form. The tickler coil was wound on the same form as the main tuning inductor (far right coil in photo Figure 7-9).

Plug in coil forms are no longer manufactured. Some sources suggest using the plastic base from a defective vacuum tube as a coil form, but tube bases are also in short supply. Nine pin computer D-Sub plugs and jacks are widely available today and can be used as a base for a regenerative radio plug in coil. A matching D-Sub jack can be mounted on the radio chassis or breadboard. The capacitance of each pin of the D-Sub jack and plug is about 7 pF to chassis ground. This has to be compensated for in the receiver design.

Switches in RF circuits are notoriously ill behaved over time. The contacts get dirty, causing noisy band switching and sometimes total inoperation of the radio. Plug in coils can also have dirty contacts and can malfunction, but they are generally less problematic than switches. The fewer terminals a coil has to be switched, the less likely a problem with switching will occur.

Many of the receiver designs in this book use simpler two terminal inductors. These are often hand wound powdered iron or ferrite core toroids (similar to Figure 7-9, far left). Two terminal toroidal inductors can be switched with a simple one pole rotary switch or a PC board dip switch. Although cumbersome to operate, the dip switch band switch used in Figure 11-5 is very reliable. The Colpitts circuit is the only configuration that utilizes two terminal coils.

Commercial off-the-shelf two terminal solenoid inductors such as the J. W. Miller line, have a lower Q than the handmade toroids described in Table 1-2. But commercial solenoids can be used in tuning circuits. Table 1-1 shows a list of tuning inductor values that can be used with a nominal tuning capacitance of 113 pF on the various shortwave bands with most of the radio circuits in this book.

Panels

The panel is the vertical board or metal plate attached to the front edge of a breadboard or chassis and used to mount the controls and dials of the radio. The hobbyist will usually construct his own panel. As with circuit boards, there are now Internet services that will create metal panels to your specification for about fifty dollars.

Although metal panels provide better shielding against the capacitive effect of the human body, nonmetallic panels are usually adequate for most radios.

The panel artwork can be created using a computer drawing program (see Figure 11-9). The artwork can be printed on pastel colored paper, which can then be laminated. Glue the laminated paper to a thin particle board, then cover the edges with poster edging. This makes a professional looking panel (Figure 11-6A).

I have tried using toner transfer media to transfer computer generated artwork to a metal panel. I have not been successful in obtaining a good panel by this method. The method requires the application of uniform heat to transfer media. The metal panel tends to heat unevenly.

Frequency Dials

Until the late 1920s, most commercial radio receivers did not have calibrated frequency dials. A log type of tuning dial (Figure 4-5) was marked with the numbers 0 to 100. When a station was located, the log dial setting was recorded instead of the actual frequency of the station so that the station could be located in the future. TRF radios had three separate dials that had to be set independently to tune in a station.

For homebuilt receivers, it is necessary to make a separate frequency dial for each individual radio. That is because the characteristics of the tuned circuit are affected by the particular inductors, varactors or variable capacitors used for each radio.

If designing a kit, and a large number of identical shaft operated variable capacitors could be obtained, then a calibrated dial could be supplied with the kit. After a warm-up period, the radio could be calibrated at one point on the dial. The rest of the dial would then be fairly accurate. But identical air-variable capacitors cannot be obtained.

In the case where the frequency range covered is large with respect to the center frequency, the tuning characteristic of the frequency dial will be exponential. The stations will be compressed close together at the high frequency end of the dial. On the medium wave broadcast band, this effect can only be partially compensated for by using a nonlinear tuning capacitor. Analog tuning dials on medium wave receivers are always nonlinear. Absolute frequency linearity in a receiver can only be achieved with a digitally synthesized local oscillator.

When a single shortwave band is tuned with one rotation of a linear variable capacitor, the tuning will be almost linear. That is because the change in capacitance over the band is small with respect to the total capacitance of the tuning circuit. Table 9-1 demonstrates this numerically. This method of obtaining linearity is widely used in this book. The method requires that each shortwave band have a separate tuning inductor.

A nine band dial that was made for a regenerative radio is shown in Figure 7-10. This dial was for use with a mechanical variable capacitor that

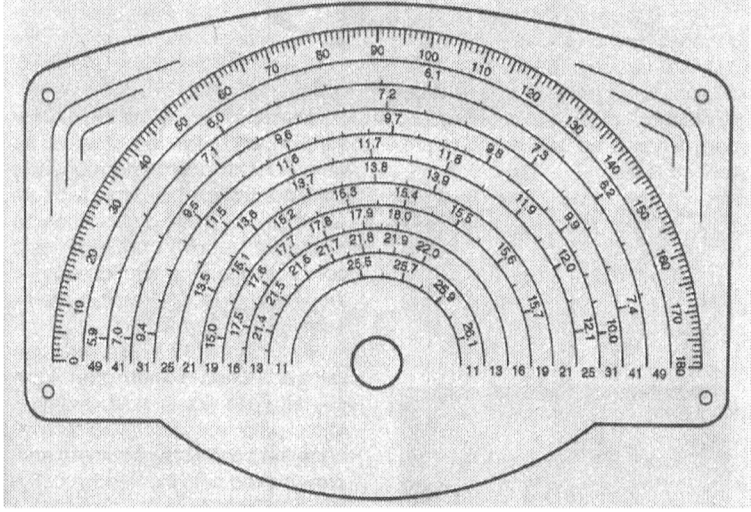

Figure 7-10: Tuning Dial For a Nine Band Variable Capacitor Tuned Receiver

produced nearly linear tuning. Figure 7-11 shows a dial prepared for a six band radio that is varactor tuned. The tuning voltage was supplied by a potentiometer. That is why the dial in Figure 7-11 covers a larger angle than the 180 degrees of the dial in Figure 7-10. The potentiometer is an audio type which has a logarithmic curve that compensates somewhat for the exponential curve of the varactors. However, the frequency spacing is stretched out near the center of the dial.

The dial in Figure 7-11 is similar to the manufactured dial in Figure 9-1.

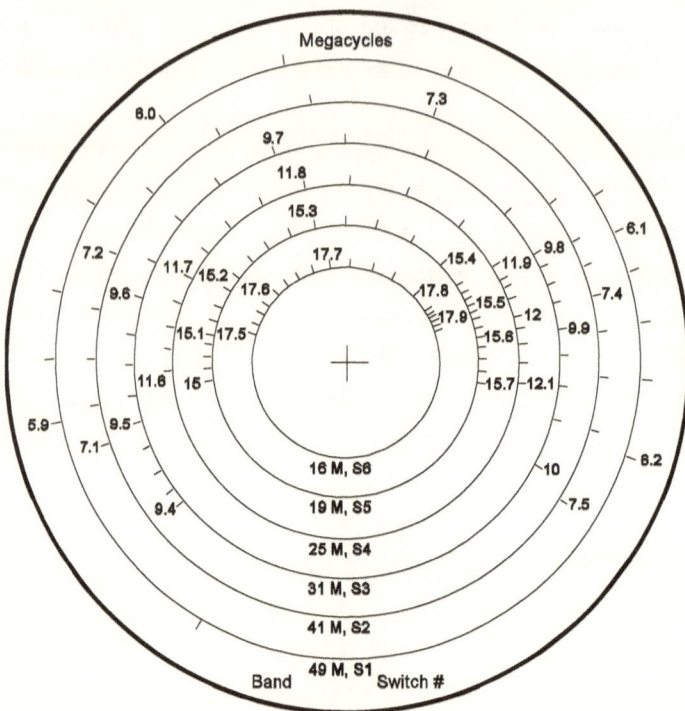

Figure 7-11: Tuning Dial For a Six Band Varactor Tuned Receiver

In the latter case, the approximately 300 degree tuning angle is produced with a 180 degree variable capacitor by the use of gears.

To make a frequency dial for a specific receiver, draw concentric circles on a piece of paper and center the paper under the pointer of the tuning capacitor or potentiometer. A frequency counter can be used with an RF generator to produce accurate frequency steps. Tune the radio in equal frequency steps and mark the frequency locations on the paper dial.

Another technique of making dials is to connect a frequency counter to the radio frequency output of a regenerative radio and turn up the regeneration control until the radio oscillates. The frequency will then be displayed on the frequency counter. This method is less accurate than the method above because the frequency of the radio is slightly different in the receiving mode than in the oscillating mode.

The fastest method of calibrating the dial once a single frequency is located is to use a marker generator (Figure 10-8). A marker generator produces a square wave at a precise frequency such as 25 kHz. This signal is coupled to the radio antenna. The regeneration is set for receiving. The radio will indicate harmonics at 25 kHz intervals across the band. These harmonic frequencies can be quickly marked on a paper dial. Since one frequency on the dial is known, the frequency of all the other harmonics can be readily deduced.

The harmonic frequencies from a marker generator are not modulated and are heard on a regenerative radio as a hissing sound or a hum. The marker generator is most accurately used on a superhet radio having a tuning meter.

Knobs and Pointers

A number of control knobs are shown in Figure 7-12A. The underside of the same knobs is shown in Figure 7-12B. The knobs nearer the top of the photos are the oldest. The most common shaft size for potentiometers and variable capacitors is one quarter inch. Some variable capacitors use smaller shafts such as 3/16 inch. Adaptors have to be used on these shafts if they are to be used with 1/4 inch knobs or couplers. Also, knobs for smaller shafts can sometimes be found at surplus outlets.

Plastic knobs with decorative metal inserts such as those that became popular in the 1950s (lower right of Figure 7-12A) are still manufactured today. Older style plastic knobs without inserts are rare. Plastic pointer knobs that are popular for instruments are still available essentially unchanged from the original appearance (lower left corner and center of Figure 7-12A). Pointer knobs are a good choice for many radio projects where an old fashioned appearance is desired. Wooden control knobs (top center of Figure 7-12A) are very rare today, but could possibly be constructed from wooden cabinet knobs.

Large dial pointers are virtually unobtainable and have to be home constructed. One method of pointer construction is to straighten the wire from a large paper clip to form a pointer. The pointer wire is then attached to a common plastic control knob using strong glue or epoxy. The wire can be painted black if desired. Instead of a wire, the pointer can be made from a thin piece of clear plastic with a straight line scored on the bottom surface (Figure 11-6A). The plastic pointer then must be glued to a common control knob.

Figure 7-12A: Control Knobs

Figure 7-12B: Underside or Above Control Knobs

Dial Lighting

Antique radios often had frequency dials that were back lighted or front lighted. These radios can be operated and the frequency setting could be observed under any conditions of room lighting, including total darkness. Backlighting was accomplished by small incandescent bulbs behind a translucent printed frequency dial.

Sometimes, the frequency dial was front lighted. The dial was opaque and the incandescent bulbs were placed between the dial and the cabinet front to illuminate the front surface of the dial. With this method, it is necessary to recess the dial somewhat behind the surface of the front panel to make room for the bulbs.

There exists the possibility of using light emitting diodes (LEDs) for dial backlighting instead of incandescent bulbs. A number of LEDs would have to be arranged in an array behind a translucent dial. The angle at which light is emitted from a single LED is limited, but wider angle LEDs are becoming available. This light source is more difficult to construct than incandescent lighting, but a multi-LED light source is more uniform and the power consumption of an LED array will be less than incandescent bulbs.

The LED assembly will last indefinitely without replacement. Green LEDs are cheap, readily available, and would be a good choice for dial lighting. Red LEDs are more common and produce more light than green LEDs, but the red color would be unpleasant for dial lighting. White LEDs are becoming available, but they are currently expensive. The price of white LEDs will decrease as they begin to be manufactured in large quantities.

8 Vacuum Tube Radio Design and Components

Availability of Parts

Performance of a regenerative radio designed with vacuum tubes is about the same as a semiconductor design. But a vacuum tube radio will be considerably more expensive to build than the semiconductor equivalent. Also, some parts will be more difficult to obtain. The extra cost is due to the cost of tube sockets, the tubes themselves, the extra and more expensive transformers, and the metal chassis, if used.

The transformers and most other parts for tube projects may be purchased from electronics parts distributors or Radio Shack stores. The items not available from ordinary electronics mail order sources are vacuum tubes, tube sockets, and variable tuning capacitors. New tubes are still manufactured in Eastern Europe and Russia, but the Internet has many suppliers of surplus tubes and this is the most economical source at present. There are specialty suppliers (such as Antique Electronic Supply) from which original tube circuit parts can be obtained.

As mentioned in the last chapter, high quality variable tuning capacitors, which are also needed for transistor circuits, are one of the most difficult items to obtain and availability changes over time from one surplus supplier to another. An alternative to the mechanical variable capacitor, which is discussed later, is varactor tuning diodes with which the capacitance is changed by a variable DC voltage. Another alternative is for the homebuilder to construct his own variable capacitor.

Vacuum Tubes

Early tubes such as the 01A and the WD11 that were used in the first tube receivers are no longer available. Non-working tubes of this type are collectors items. If these early tubes were available, I would be reluctant to use them because of their unpredictable characteristics and low gain. Due to a poor vacuum in the early tubes, there was considerable grid leakage and a grid resistor often did not have to be included in the circuit. Grid resistors were required when high vacuum tubes were developed.

I use tube types from the 30s, 40s and 50s, which can be purchased for reasonable prices today. The indirectly heated cathode tubes are easier to use because the filament power can be AC supplied by a step down transformer. The filament circuit is not shown in the tube drawings of this book. Six or twelve volt tubes are suggested and AC or DC power can be used for the cathode heater supply. The many specialty tubes of the 1960s, such as the compactron, were not manufactured in large numbers and are rare today.

Early Dry Battery Tubes

The base of the early battery type triodes was a plug with four equally spaced pins arranged in a square pattern. Two pins were larger than the other two. This prevented the tube from being plugged in the socket in the wrong orientation.

These tubes were made to be powered by dry batteries. When in use, the filaments of these battery tubes did not light up and did not give off appreciable heat. In that respect, battery tubes are as visually undramatic as transistors. After the advent of indirectly heated cathode tubes, battery tubes continued to be manufactured with octal and miniature bases for use in portable battery operated radios. These later battery tubes are more rare than indirectly heated cathode tubes of the same era because they were manufactured in smaller quantities and fewer surplus tubes are now available.

Indirectly Heated Cathode Tubes

The indirectly heated cathode tubes contain a filament that gives off an orange glow and the tube gets warm while in operation. This lends to the exotic quality of vacuum tube equipment and this visual quality is even more evident when the receiver is operated in a darkened room. If the tube is made of glass, the glowing filament inside the tube can usually be seen. However, in many glass tubes, the filament is partially obscured by a silver coating inside the glass which is formed during manufacturing by firing a "getter". The getter absorbs gas inside the tube and enhances the vacuum. The indirectly heated cathode tubes are the cheapest and most available types today. The filament consumes too much power for use with portable dry cell radios. Although developed for AC operated radios, these tubes were also used in six volt and twelve volt automobile radios, which were operated from a car DC electrical system.

The first class of tubes that had a uniform numbering system were used in the early 1920s and had part numbers such as UX-242 or UY-242. Later in the 1920s, the UX or UY was dropped and only the number 242 was used. Later yet, the leading 2 on the tube number was dropped leaving the type number 42. The base of this class of tubes was a plug with four to seven pins arranged in a circle. With some of the base types, two of the pins were larger than the others to prevent misalignment. In other types the pins were the same size, but were at unequal spacing on the circle of the base thus preventing misalignment.

This class of tubes were intended for AC line operation. Some of them had indirectly heated cathodes and some were directly heated. In the later case, various techniques were used to reduce the hum generated by the AC filament supply. Filament voltages ranged from 2.5 to about 7.5 volts. These tubes are fairly rare and expensive today. The variety of filament voltages of this class of tubes leads to inconvenience in design.

The octal tube has eight pins equally spaced in a circle and a center key to prevent the tube from being plugged in the wrong way. These tubes appeared in the 1930s and had names such as 6V6. The first number of the part name indicated the voltage of the filament, in this case 6.3 volts.

The middle letters of tube type numbers were chosen sequentially as new tube types were developed. The letters X, Y, and Z were reserved for rectifier tubes. When the single alphabet letters were used up, then double letters (i.e. 6AB7) were used. One group of octal tubes, such as the 6J7, had a cap at the top of the tube, which was usually the control grid connection. The rationale of this design was to reduce the internal capacitance of the grid to a minimum. A later modification of these tubes placed the grid connection at the base of the tube with the other connections. The part number of these modified tubes contained an "S" in front of the original designated letter (i.e. 6SJ7).

The last number in the tube name indicated the approximate total number of elements in the tube. There was a specified way of counting tube elements which was not consistently adhered to.

The indirectly heated cathode tubes worked the same on automobile DC supplies as on AC. Of course, plain filament tubes would have worked as well on automobile DC power. But as indirectly heated cathode tubes were manufactured in large numbers for home AC equipment, their price was lower than a special automobile radio tube would have been. Dry battery tubes were manufactured in the octal and miniature styles for use in dry battery operated portable receivers. These tubes had numbers that started with 1 or 3 for filament voltages of 1.5 volts or 3 volts (i.e. 1N6).

Figure 8-1 shows the outline drawings and photographs of several types of tubes. Octal tubes were available in three different styles G, GT, and metal. The GT style was glass with a cylindrical (tubular) shape. The term GT meant "glass tubular". The G (glass) type tubes featured a glass cylindrical shape with

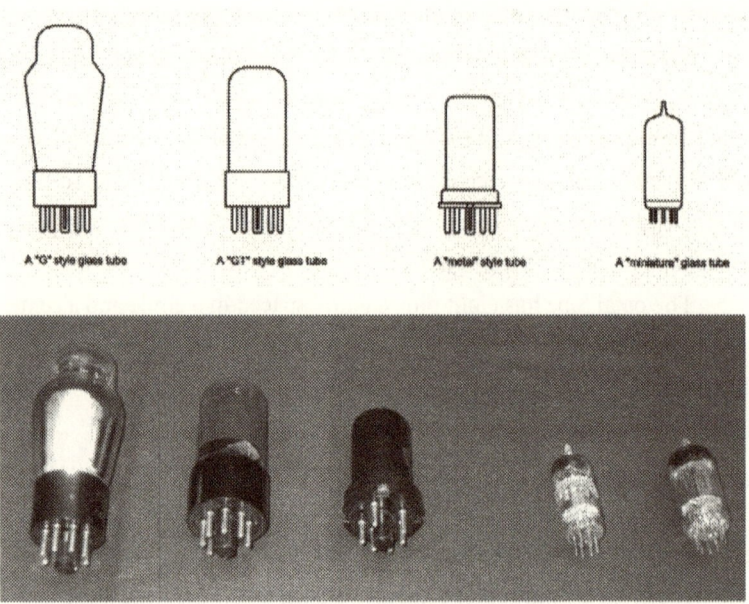

Figure 8-1: Octal Base and Miniature Glass Vacuum Tubes

a bulge in the middle of the cylinder. The G style tubes were manufactured earlier than the GT style and are rarer today. The metal tube was a cylindrical glass tube with a metal cover, which provided shielding for the tube. The metal tube had no suffix to the part number. The G, GT, and metal tube styles with the same tube numbers (i.e. the 6J5G, 6J5GT and 6J5) are essentially interchangeable. As mentioned earlier, the filament inside of a metal octal tube cannot be seen due to the metal cover. The metal tube does get warm after a few minutes of use.

Miniature tubes became available in the late 1940s, but did not become popular until the 1950s. The tubes are all glass and the plug is built into the bottom of the glass (last drawing of Figure 8-1). Miniature tubes are smaller than octal tubes. The plug consists of 7 or 9 pins arranged in a semicircle. Both types are shown in the photo. A larger gap between two of the pins prevents the tube from being inserted with the wrong orientation. Miniature tubes share the same numbering system as metal octal tubes. There is no G or GT suffix. Most miniature tubes have double letter designations (i.e. 6AU6) because they were numbered after the octal tubes.

However, it cannot be definitely determined whether a tube is octal or miniature by the tube number. For instance, 6AB7 sounds like a miniature tube number, but is actually a metal octal tube. Automobile systems went to 12 volts in the 1950s and 12 volt tubes (such as 12BA6) began to appear. Twelve volt tubes also became popular in AC series string radios and televisions. Some twelve volt types such as the 12AX7 have a center tapped filament which can be operated on either six or twelve volts. Figure 8-1 shows the outline drawings and photos of the three octal tubes and a miniature tube.

Octal and miniature tubes are still available, usually at reasonable prices and they can be used for new designs. The octal tubes are more exotic looking, but the miniature tubes are all glass, they are available in more configurations, and tend to be cheaper. Miniature and octal tubes were often mixed in the same circuit (especially in televisions) in commercial designs of the past.

Rectifier Tubes and Diodes

One type of tube used in older equipment was the rectifier tube. This tube consisted of one or two power diodes and was part of the DC power supply. The tube was used to convert the AC line voltage to pulsating DC. The pulsating DC was then filtered to produce pure DC.

High voltage silicon rectifiers have been available for 40 years and the rectifier tube can be replaced by them. The silicon diodes can be placed inconspicuously below the chassis. Silicon diodes will have a lower forward voltage drop than a rectifier tube and will result in a better regulated power supply. Also, the diodes will last indefinitely whereas a rectifier tube will not. For these reasons, I use silicon rectifiers in new tube designs instead of ratifier tubes.

One cannot replace a rectifier tube with silicon diodes in older equipment without some redesign. Replacing a rectifier tube with silicon diodes will cause the DC voltage to increase by about 20 volts. This voltage rise could damage tubes and filter capacitors if not compensated. There is a new type of silicon diode called the Schottky diode that has a lower forward drop than conventional silicon diodes. The forward voltage drop is about 0.5 volts instead of 0.7 volts. For tube equipment, the Schottky diode offers no significant advantage over regular diodes.

Transformers for Tube Designs

One of the major costs in tube designs is the required transformers: high voltage power, filament power, and audio output transformers. There are a few sources for original tube transformers on the Internet but the prices are high. The power transformer which had a high voltage winding for the plate supply can be simulated by wiring two 6.3 VAC or 12.6 VAC transformers back to back. The tube filaments can be driven from the 6 or 12 volt output of the first transformer and the second transformer can step the 6 or 12 VAC back up to 117 VAC. This voltage can be rectified and filtered for the plate supplies (see Figure 11-10).

For the audio output transformer, a multi-output power transformer such as the Mouser 41FW300 can be used. The 220 volt winding is used as the primary. The nine volt output winding is about the right turns ratio for a 6V6 or 6AQ5 tube operating at 120 VDC with an eight ohm speaker load. The audio output power of a single ended power amplifier is limited to about 450 mW into 8 ohms with a 6V6 or 6AQ5 tube. This output power is adequate for a shortwave receiver. The fact that a 60 cycle power transformer can be used as a speaker output transformer is surprising. Theoretically, the DC plate current of the output tube would cause the transformer core, which is not designed for DC use, to saturate. However, my experience is that the transformer core under these conditions does not saturate and the power transformer works fine as a speaker transformer.

If the plate supply voltage of the 6V6 tube were increased to 320 volts, the power output of the tube would increase to three watts. With a 50C5 or 35C5 tube used with 120 volts supply, over one watt of audio output power can be obtained. But an isolated power source for the 35 or 50 volt filament would have to be found. Either alternative for increasing the audio output power would complicate the design.

AC-DC Series String Radios

A method used for eliminating the power transformer in tube equipment was the AC-DC series string filament design. The tubes for this design had a high voltage filament between 12 and 50 volts instead of the usual 6 volts when a power transformer was used.

A typical tube number for an AC-DC tube is 50L6-GT as opposed to 6L6 for a tube for an isolated power supply. The filament of the AC-DC tubes were selected so that the filament voltages added up to somewhat less than 117 volts. Each tube filament had to be rated for the same current. The tube filaments were connected in series with a power resistor added to the series string. This string is connected across the 117 volt AC line. Thus, no transformer was needed for the filaments.

The plate supply voltage was obtained by half-wave rectification of the 117 volt AC line voltage. The tubes were designed to work on the resulting DC supply voltage of around 110 volts. Because there was no power transformer, the ground of the radio was necessarily connected to one side of the AC power line. This resulted in a shock hazard if any metal part of the radio was touched while touching an electrical ground at the same time. If the chassis of such a radio was touched while standing barefoot on a cement floor, a shock would be felt.

The AC-DC radios were usually totally encased in a plastic box in order to prevent operator contact with the power line connected chassis. Plastic control knobs prevented contact with the control shafts. But often there was an exposed screw somewhere on the back of the radio that could deliver a nasty shock. As late as 1950, some large cities in the USA still had a 110 volt DC power distribution system instead of an AC system. The AC-DC radio would work with a DC power line supply although the power plug had to be properly inserted to provide the correct DC polarity to the radio. This explains the name AC-DC which was applied to this class of radios that were not isolated from the power line by a transformer.

Many TV sets used the series string filament concept, although due to the larger number of tubes, six volt tubes could be used in the series filament string. One big disadvantage to the un-isolated design was that the shock hazard was extended to any device that might be electrically connected to the radio or TV. This includes phonographs, microphones, external speakers, headphones, antennas, and test equipment. Repair technicians had isolation transformers to place between the radio and power line so that the sets could be safely repaired.

The power transformer radio preceded the AC-DC design. The isolation provided by the power transformer in old radios was sometimes defeated by adding capacitors and resistors between the chassis and the power line. The reason these parts were added was to utilize the power line as a ground connection for the radio. Even without these added components, the radio chassis is coupled to the power line ground to some extent through the stray capacitance between primary and secondary of the power transformer. This coupling may be sufficient to provide an RF ground, but the stray capacitance of a transformer is not great enough to cause a shock hazard.

Although the power line can provide an effective ground for a radio at medium wave and shortwave frequencies, the power line contains a great deal of noise that is likely to degrade the reception. In modern designs, the coupling to the power line through the power transformer capacitance is often broken by RF inductors in series with the power line. A similar effect can be achieved by encircling the power cord with a ferrite core. The radio can then be attached to a clean ground such as a metal cold water pipe or metal rod driven into the earth nearby the radio. If an antenna with a long coaxial feed is used, the shield on the coaxial cable often serves as an effective ground.

The Ocean Hopper tube radio kit that will be described in Chapter 11 was an un-isolated design on an open metal chassis. To lower the shock hazard the chassis was connected to the power ground through a parallel resistor and capacitor. This was hardly effective and frequent shocks were experienced while using the radio. The headphone connection of this radio was to the plate of the power output tube though a pair of capacitors. As the capacitors had considerable DC leakage, it was not uncommon when wearing headphones to get a shock to the ears! To make matters worse, the sides of the metal chassis were open so that fingers could stray into the wiring contacting AC and DC voltages over 100 volts.

The non-isolated AC-DC design should never be used for new radio designs. It is highly likely that hobbyists will want to connect external devices and test equipment to their radios without creating a shock hazard. A power transformer is always desirable in new designs in spite of the extra cost. For the most part, the high filament voltage type tubes intended for the AC-DC radios (such as the 50C5), though available, will not be used by hobbyists except for replacement in existing radios. There is no reason that these high voltage filament tubes could not be used in transformer designs if an isolated filament voltage can be supplied for them.

9 Semiconductor Radio Design and Components

Bipolar and Field Effect Transistors

Discrete bipolar transistors, JFETs, and MOSFETs are still widely available today. Individual part numbers of discrete transistors are sometimes discontinued, but substitute transistors are always available. There is no apparent trend to discontinue the discrete class of semiconductors.

The 2N3904 NPN bipolar transistor has a gain of 100 and is useable up to 30 mHz, the highest frequency of the shortwave range. This transistor comes in a small epoxy case and is good for most receiver design applications.

The 2N5486 junction field effect transistor (JFET) is likewise useful for shortwave applications. The transistor has a fairly high transconductance (gain) and is usable into the ultra high frequency (UHF) region.

The metal, oxide, semiconductor field effect transistor (MOSFET) is similar to the JFET in operation, but is created by a totally different process. The device has a very high transconductance and has been used for this reason when the extra gate was not needed.

As discussed previously, the dual gate MOSFETs were unavailable at reasonable prices for several years. Fortunately, the function of this part can be duplicated by using two JFETs. A regenerative receiver using the dual gate MOSFET is shown in Figure 5-10 and the equivalent JFET radio is shown in Figure 5-11.

Integrated Circuits

Integrated circuit (IC) designs are often discontinued after a few years leaving the hobbyist with no replacement for previous designs. One example of a discontinued IC is the MC3340P electronic volume control IC that I once used in a radio design. Electronic volume controls are now included in audio integrated circuits such as pre-amplifier ICs. But using a substitute part requires redesigning the old PC board.

Digital potentiometer ICs have replaced some of the functions once performed by voltage controlled volume controls like the MC3340P. These ICs are now available to hobbyists and can be used as volume controls that are set with two push button switches. One switch moves the electronic slider up the internal resistance and the other switch moves the slider down. The IC potentiometer can also be controlled by a digital rotary encoder. When this is done, the control works similar to a conventional analog potentiometer. Some commercial automobile receivers use this technique for volume control.

The digital potentiometer can be visualized as a large string of series resistors with an electronic switch that can connect between any two of the resistors in the string. Because digital potentiometers change values in discrete steps, they cannot be used for varactor tuning or for regeneration controls. These applications require a continuously adjustable control. One advantage of digital potentiometers is that the control is precise and the two halves of stereo potentiometers track very closely. This means of volume control is much more complex than the old MC3340P IC and probably would not be used in a home-built receiver unless remote control of the receiver was desired.

Because of specific IC designs becoming obsolete, I try to minimize the number of ICs in my designs. Hobbyists electronic circuits are often built years after the design is first published. Designs of manufactured electronic products are used for about six months to three years before a new design is required. Therefore, a manufactured design is discarded long before the constituent ICs become obsolete and unavailable.

One IC that I do use and that has survived for about 30 years is the LM386 power audio amplifier. It seems likely that if this part is discontinued, there will be similar ICs available with the same pin configuration that can be directly substituted. The 555 timer IC has also been around for several decades and seems likely to be available in one of its versions for a long time.

Operational amplifiers (op amps) keep coming out in improved designs, some with the same pin configuration as older designs. Op amps were originally useful only at DC and audio frequencies. Increases in bandwidth and slew rate of modern op-amps allow them to be used at radio frequencies (RF). This makes possible new forms of RF filters, variable gain amplifiers, and precision rectifiers. The latter circuit can be used as an AM detector. Variable gain RF op amps can be used as AM modulators. An op amp is used in Figure 9-3.

Passive Components

For tuning circuits, silver mica capacitors are preferred for their stability and low loss. These capacitors are rated at around 500 volts DC so they can be used for tube circuits as well as for semiconductor circuits. Tantalum or the new high capacity ceramic capacitors are preferred for higher capacitance applications due to the low equivalent series resistance (ESR) of these types.

Metal film resistors are preferred for RF circuits because of their low parallel capacitance. For the audio and power section of a receiver, standard carbon resistors, ceramic disk capacitors, and aluminum electrolytics can be used.

The point contact germanium diode, because of its low forward voltage drop, has been traditionally used as an AM detector. The newer hot carrier diodes such as the MBD101 have slightly more forward voltage drop than germanium diodes, but can be used to replace them. One AM detector design of this book uses a technique to cancel the forward voltage drop of an ordinary silicon diode such as the 1N914 so that this diode can be used for AM detection.

Frequency Linearity of Tuning Circuits

The frequency versus capacitance characteristic of a tuned circuit over a wide frequency range is nonlinear. On the medium wave broadcast band (530 to 1710 kHz), the bandwidth is 120 percent of the center frequency of the band. The dials of nearly all analog medium wave receivers are nonlinear with the received frequencies bunched together at the high frequency end of the band. This occurs even when the variable capacitor is designed to compensate for the effect.

Figure 9-1 shows the dial of an old commercial receiver. Even though the variable capacitor is compensated, the frequencies are still compressed together on the high frequency part of the dial. Notice in Figure 9-1 that the tuning angle of the 100 kHz space between 550 kHz and 650 kHz is about three times as wide as the 100 kHz space between 1600 kHz and 1700 kHz. The shortwave portion of this radio covers the frequencies of 5.5 mHz to 19 mHz . This frequency space contains seven international shortwave broadcast bands and

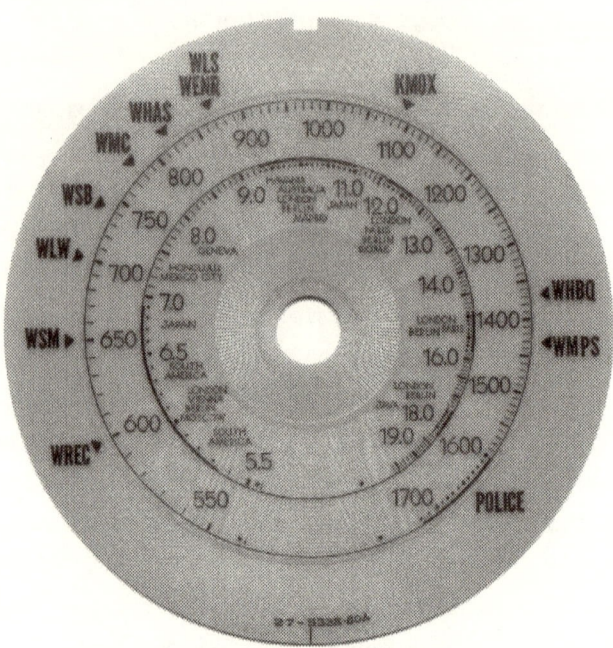

Figure 9-1: Dial From a 1936 Philco Radio

every frequency in-between. Five of the international bands are indicated on the dial by boldface dots replacing dial lines. The frequency compression at the high frequency end of the shortwave section of the dial is even more pronounced than the compression in the medium wave section.

On the single 49 meter short wave band, (5950 to 6250 kHz) the width of the frequencies covered is only 5 percent of the center frequency. Table 9-1 is a spread sheet calculation that shows the change in tuning capacitance necessary to produce equal frequency steps in the circuit shown above the table. Column 4 of Table 9-1 shows the percentage capacitance change with respect to the total tuning capacitance at that frequency. Column five shows the percentage change of tuning capacitance with respect to the total tuning capacitance change over the entire band.

The changes in total capacitance shown in Table 9-1 vary a little more than one percent over the entire 49 meter band. In other words, on a dial covering just one international shortwave band, the frequency versus capacitance curve for a tuning circuit is nearly linear.

Mechanical linear variable tuning capacitors produce a nearly linear

TABLE 9-1

Changes in Tuning Capacitance Needed to Produce Equal Changes in Frequency

Frequency KHz	Capacitance Total, pF	Capacitance Change, pF	% Change of Total Cap vs Capacitance	Capacitance % Change vs Total Change
5950	152.39			
5975	151.12	1.27	.835	8.91
6000	149.86	1.26	.831	8.80
6025	148.62	1.24	.828	8.69
6050	147.39	1.23	.825	8.58
6075	146.18	1.21	.821	8.48
6100	144.99	1.20	.818	8.37
6125	143.81	1.18	.814	8.27
6150	142.64	1.17	.811	8.17
6175	141.49	1.15	.808	8.07
6200	140.35	1.14	.805	7.97
6225	139.22	1.13	.8.2	7.87
6250	138.11	1.11	.798	7.78
Total Capacitance Change, pF		14.28		

capacitance versus shaft position characteristic. The frequency versus rotation angle curve for a tuned circuit using a mechanical 4 to 20 pF variable capacitor in parallel with a 130 pF fixed capacitor on the 49 meter shortwave band is nearly a straight line. This is evident from the 9 band shortwave tuning dial of Figure 7-10.

Varactor Tuning Circuits

Figure 9-2 shows various tuning curves obtainable with varactor tuning and the tuning circuits that produce the curves. The desired frequency versus rotation curve is the straight line shown in the far right hand curve of Figure 9-2D.

The use of a potentiometer and varactors instead of a variable capacitor as the tuning device allows for tuning over a shaft rotation angle of 300 degrees compared to the 180 degree rotation of a mechanical variable capacitor. Compare the dial of Figure 7-11 with the dial of Figure 7-10. This extra tuning angle of a potentiometer has the same effect on tuning as adding a mechanical gear mechanism to a mechanical variable capacitor. The antique Philco radio, which contains the dial in Figure 9-1, has a gear mechanism to convert the 180 degree tuning sweep of the mechanical variable tuning capacitor to about 300 degrees on the dial.

Figure 9-2A shows a varactor diode circuit that can replace a mechanical variable capacitor. The potentiometer used in this varactor circuit is the linear type. The parts for this circuit are readily available. Varactors produce a logarithmic change in capacitance (and therefore tuned frequency) with respect to applied voltage. A linear change such as that provided by mechanical linear variable capacitors is desired. The nonlinearity of the varactor frequency curve is shown in the far right of Figure 9-2A.

In Figure 9-2B, the voltage supplied to the varactors comes from an audio taper or log potentiometer. The voltage versus rotation curve should be exponential as shown by the dashed line in the center curve. But audio potentiometer manufacturers approximate the exponential by two intersecting straight lines as shown by the solid lines in the center curve of Figure 9-2B.

This approximation works alright in volume control applications, but not as well with varactor tuning devices. The frequency versus shaft position in the far right curve of Figure 9-2B shows a marked nonlinearity at the center of the tuning curve. This nonlinearity of tuning near the dial center is evident in the frequency dial of Figure 7-11. The frequencies are spread out in the center of the dial with respect to the spacing at the ends.

Figure 9-2C shows a varactor circuit in which the wiper on a linear potentiometer is shunted by a fixed resistor with a value equal to about one quarter of the total value of the control. The voltage versus rotation shown in the center curve of Figure 9-2C is similar in shape to an exponential curve. I call this a "fake log" varactor control circuit. The frequency versus shaft rotation is shown in the far right curve of Figure 9-2C. After an initial steep rise in frequency, the curve becomes essentially straight as indicated by the dashed line. This circuit has the advantage that the component parts are readily available and the resultant tuning curve is very predictable.

A disadvantage of this circuit is that the load on the driving source changes with shaft position. The impedance of the DC voltage source of the control should be low with respect to the fixed resistor.

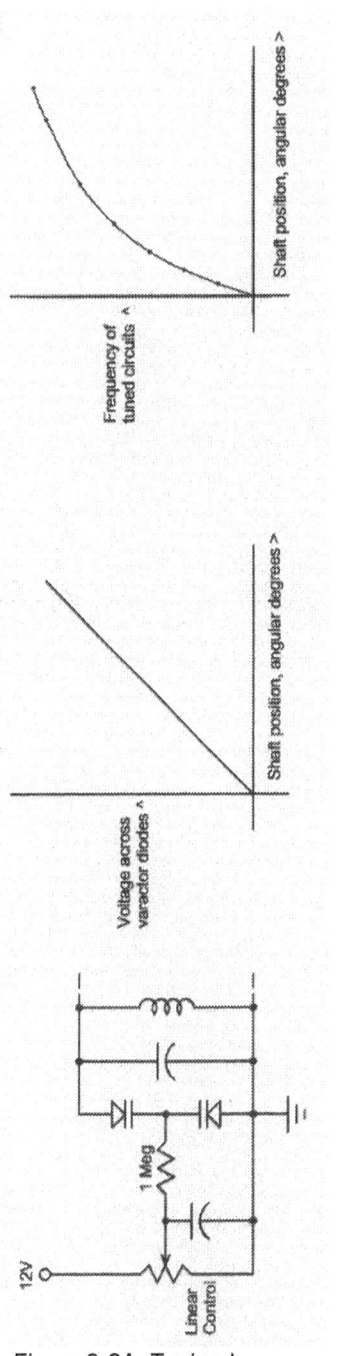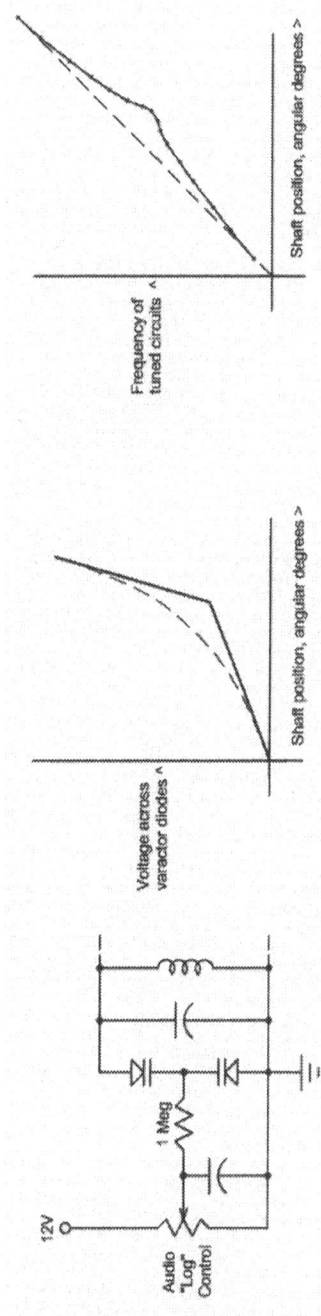

Figure 9-2A: Tuning by Varactors Supplied With a Linear Voltage Source

Figure 9-2B: Varactor Tuning With an Audio (logarithmic) Voltage Control

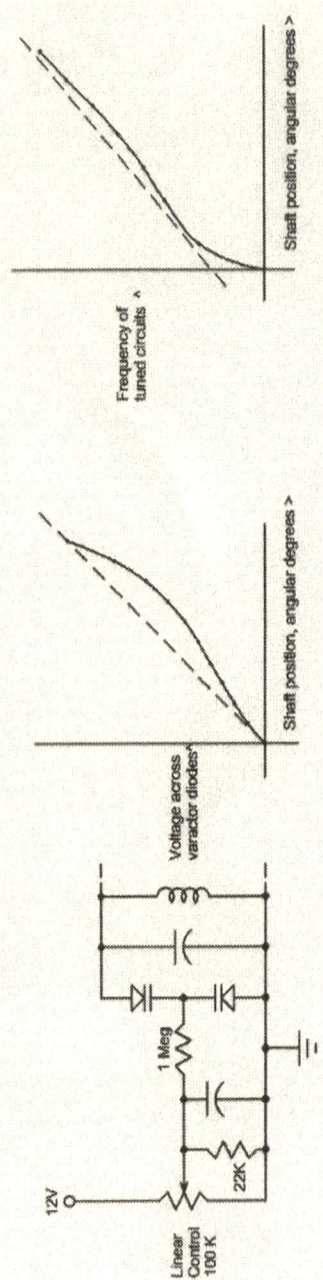

Figure 9-2C: Varactor Tuning With Voltage Supplied by a Shunted Linear Control

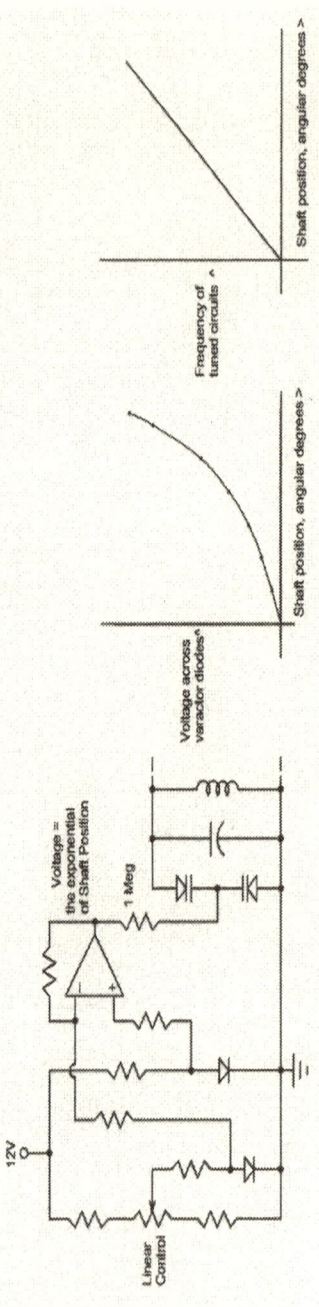

Figure 9-2D: Varactor Tuning With a Circuit That Provides Exponential Output Voltage From a Linear Control

The circuit of Figure 9-2D is a simplified version of a circuit intended to produce an exponential voltage versus shaft position curve shown in the center curve. When this voltage variation is impressed on the varactors, the frequency versus shaft position shown in the far right curve of Figure 9-2D is the desired straight line.

The circuit in 9-2D is shown in its complete form in Figure 9-3. This circuit is fairly complex and it requires some precision resistors and an operational amplifier. The linear potentiometer could be a precision servo unit. These can be obtained from surplus electronics dealers and from electronics parts sellers such as Newark. There is some drift of the circuit characteristics over time, especially when ambient temperature is not constant. Therefore, the circuit in Figure 9-3 that produces excellent frequency tuning linearity would probably not work well in an automobile environment with its wide range of temperatures.

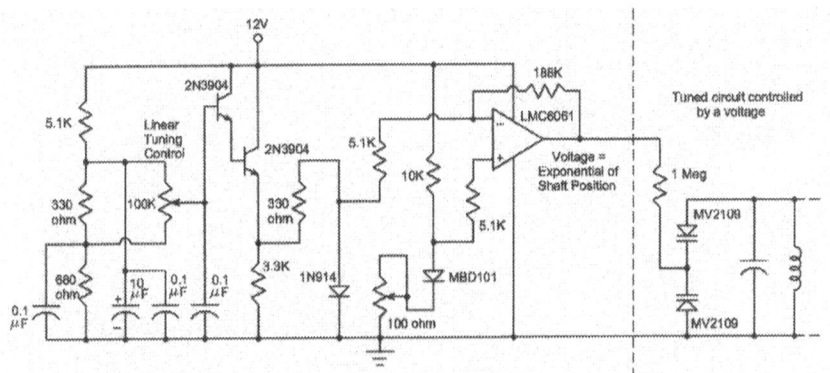

Figure 9-3: Full Details of the Simplified Circuit of Figure 9-2D

Another caveat is that the circuit of Figure 9-3 cannot completely swing the output voltage to the positive voltage rail. In Figure 11-5, it is necessary for the tuning voltage to swing all the way to the 12 volt rail in order to reach the highest desired frequency. The linear tuning curve shown in the far right curve of Figure 9-2D is an unusual accomplishment for an analog receiver.

10 Electronic Test Equipment

Commercial electronic designers use test equipment that costs tens of thousands of dollars per instrument and they maintain a special department to keep the equipment in calibration. It is particularly expensive to make precise electronic measurements. Fortunately, a lot can be accomplished in electronic design without highly precise instrumentation.

Inexpensive Testing

Many methods of relatively cheap testing have been developed by the radio amateur (ham) community. Amateur radio equipment manufacturers such as Palomar and MFJ are a source of low cost test equipment. *The Radio Amateur's Handbook* supplies information on building simple test equipment. A new edition is published yearly by the American Radio Relay League (ARRL). I purchase a new copy about every ten years.

Until around 1980, money could be saved on test equipment by purchasing "build it yourself" electronic kits. A number of my instruments are from the kit era. They are still in use because they work satisfactorily. As discussed elsewhere, the cost of manufacturing has dropped to the extent that kit building is no longer cost effective. Inexpensive instruments are now available in the fully manufactured form.

Personal Computers and Software

A personal computer is very useful for electronic design. It doesn't have to be the latest or fastest model. Software for electronics and technical drawing are available for under $150 per program. Relevant software includes general drawing programs for schematics, block diagrams, dials, and charts. A separate drawing program for creating printed circuit board patterns might be preferred. For those who like to write about their circuits, a word processor and a program to handle digital photographs is desirable.

Other useful software includes the electronic circuit simulation program. A schematic of the circuit to be investigated is drawn into the simulation program, which then calculates the performance of the circuit. Output data is displayed on virtual test instruments inside the software.

There is some controversy among engineers about the usefulness of simulation programs. They are no substitute for knowledge of electronics. The same information provided by the simulation program can be obtained by writing and solving equations for the circuits in question. But simulation programs can save considerable time when developing a new circuit.

The effectiveness of the simulation depends upon the accuracy of the mathematical models of electronic parts in the simulation program. Often, these models fail to simulate the part accurately. For instance, an incandescent lamp might be simulated as a fixed resistor, whereas a real lamp changes resistance as the voltage across it changes. This change of resistance is due to the change of temperature of the filament.

Some circuit simulation programs cost thousands of dollars, but older versions of such programs can sometimes be obtained for around one hundred and fifty dollars. Newer programs tend to expand the digital capabilities of the software, whereas the desired analog capabilities remain the same. I have a good simulation program that runs on a previous computer, but the installation is limited to one computer. I now have a new computer and I cannot install the old program. I have not found a suitable new simulation program for a reasonable price.

Pricing

The instruments discussed below, except for the oscilloscope, cost less than $150 each. A good quality new oscilloscope that will work up to 40 mHz can be purchased for around $350. Purchasing used instruments can save money. But a second hand instrument might require servicing before it can be used. Top of the line brands like Tektronics and Hewlett Packard (now called Agilent) are very expensive even when purchased used.

Older versions of software are often available at a considerable saving and the latest version is usually not significantly better than the previous one.

Some schematic drawing software and automatic PC board pattern drawing software (auto routers) are available free on the Internet from PC board prototype makers. I do not generate PC board patterns by auto routing, but prefer to draw the patterns directly.

Inexpensive Instruments

With careful shopping an entire set of new radio instruments and software can be assembled for around $1000. Equipment and software doesn't have to be purchased all at once. It can be obtained individually as it is needed. Here are some test instruments to be considered:

An analog volt-ohm-milliampere (VOM) meter is shown in Figure 10-1. A digital meter will not measure a signal that is changing at a slow rate such as one cycle per second. An analog meter, the type with a moving indicator needle, will function with a slowly changing signal and therefore should be the first voltmeter purchased. A typical price is $35.

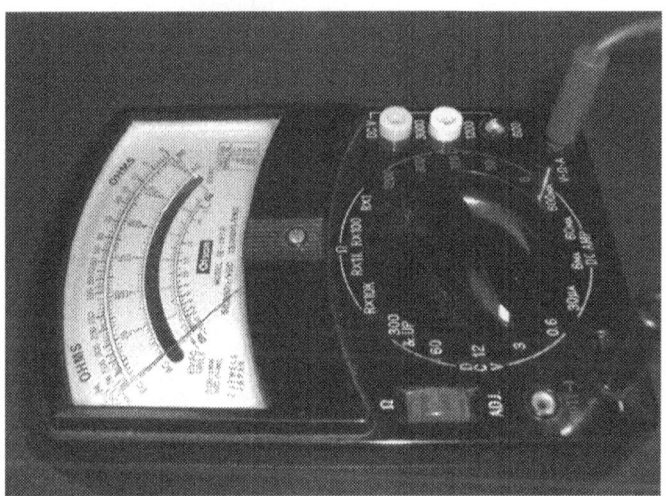

Figure 10-1: An Analog Volt-Ohm-Milliampere (VOM) Meter

A digital multi-meter, as shown in Figure 10-2, makes the same measurements as an analog VOM, but provides a numeric readout and is more convenient to use in most measuring situations. Digital meters often include a bipolar transistor "hfe" (gain) tester. An automatic shutdown feature is worth the extra money because forgetting to turn the meter off after use can waste a lot of batteries. Auto ranging meters automatically select the voltage range to be used for a given input voltage. This is a nice feature, but it is not essential. A three and one half digit meter is adequate. Some meters display more digits for greater accuracy.

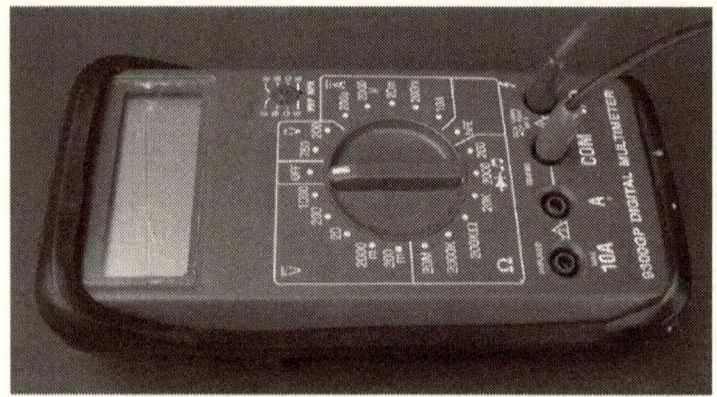

Figure 10-2: A Digital Multi-Meter

The RF diode detector circuit in the schematic of Figure 10-3 can be used to measure an AC signal with a DC meter. AC meters that will directly measure AC signals at radio frequencies are generally too expensive for hobbyists.

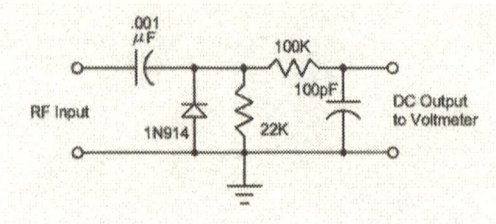

Figure 10-3: A Circuit For Measuring an RF
AC Signal on a DC Meter

True root mean square (RMS) digital voltmeters have recently become available for frequencies through the audio range. The price is around $100 depending on features. The RMS voltage is usually desired for AC measurements and the true RMS meter will measure RMS voltage regardless of the waveform of the signal. Ordinary AC meters require that the signal be a sine wave for an accurate measurement to be obtained. It is unfortunate that true RMS meters do not presently work at radio frequencies (RF).

A digital LCR meter, as shown in Figure 10-4, measures inductance, capacitance, and resistance as does an impedance bridge. At low values of inductance and capacitance, the accuracy of the LCR meter is poor, but the instrument does provide a rough estimate of the value.

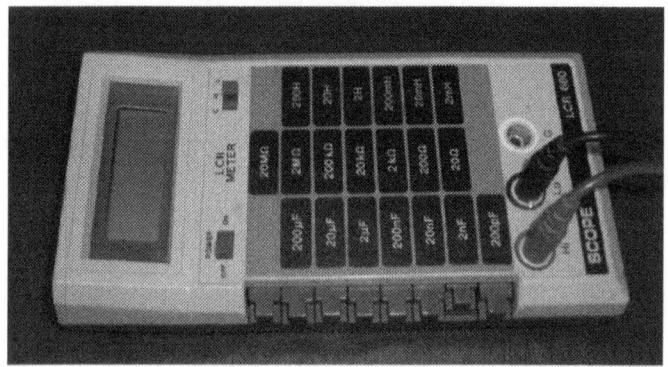

Figure 10-4: A Digital Inductance-Capacitance-Resistance (LCR) Meter

An audio signal generator that will generate sine and square wave signals of 30 Hz to 1 mHz is shown in Figure 10-5. Digital function generators accomplish the same thing and they often provide additional output wave forms such as triangle waves.

Figure 10-5: An Audio Signal Generator

A radio frequency (RF) signal generator is shown in Figure 10-6. It will produce RF signals from about 150 kHz to 112 mHz. The RF signal can be AM modulated by an internal 400 Hz signal or an external audio signal applied at the audio jack of the instrument.

Figure 10-6: A Radio Frequency (RF) Generator

An inexpensive frequency counter is shown in Figure 10-7. It measures frequency accurately by digitally counting the transitions in a signal. Inexpensive units like this one will count well into the VHF range (around 200 mHz).

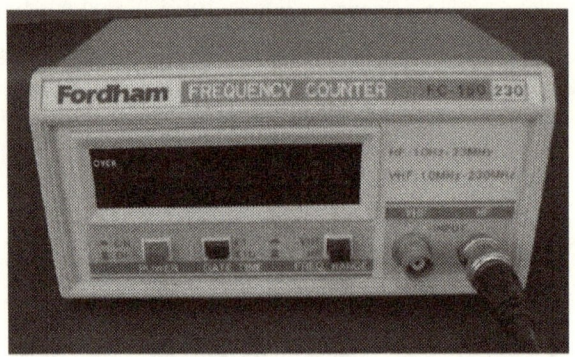

Figure 10-7: An Inexpensive Frequency Counter

A homebuilt crystal controlled marker generator is shown in Figure 10-8. This instrument is used to calibrate radio dials. A marker signal is produced at equal frequency intervals across the shortwave band that is tuned by a receiver. The ARRL Handbook (1998 edition and others) has a circuit diagram and PC board pattern for the device. This instrument works best with analog receivers that have a signal strength meter. For receivers not having a meter (such as all regenerative receivers), a modulated marker generator would be desirable as it would be possible to hear the markers as an audio tone. I have not seen any

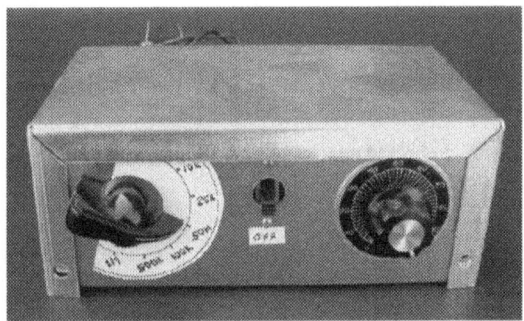

Figure 10-8: A Homebuilt Crystal Controlled Marker Generator

designs for a modulated marker generator and I have not designed one yet.

The noise impedance bridge shown in Figure 10-9 is used with a communications receiver to measure the RF impedance of antennas or tuned circuits. The receiver used with this bridge must have a coaxial antenna input, be able to tune frequencies accurately, and preferably have an analog signal strength meter. Noise bridges are available from amateur radio suppliers such as Palomar and MFJ.

Figure 10-9: A Noise Bridge For Measuring Impedance

For working with vacuum tubes, a good instrument to have is a transconductance type tube tester. I do not have one of these. I have a collection of used tubes and with a tester I could discard the ones that are not good. I could also verify that a specific tube is good before placing it in a circuit. Used tube testers can be purchased on Ebay.

An oscilloscope, as shown in Figure 10-10, is the most expensive instrument needed by electronic hobbyists. The device provides an actual picture (voltage versus time plot) of AC signals. Most all modern versions have a triggered sweep and a response down to DC, which is desirable. The higher the operating frequency of the oscilloscope, the better the instrument and the more expensive it will be. The oscilloscope pictured in Figure 10-10 has neither a triggered sweep nor response to DC and the highest frequency is only 5 mHz. The instrument is still useful with all of these limitations.

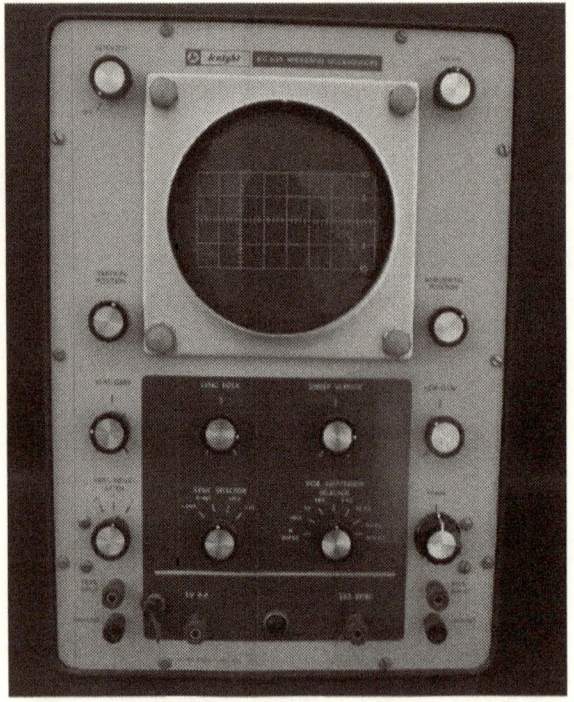

Figure 10-10: An Oscilloscope

There are plug in cards for personal computers that allow the computer to be used as an oscilloscope. Oddly, these cards are more expensive than a comparable stand-alone oscilloscope. An oscilloscope that is not tied to a personal computer is more convenient to use.

A tool that I have recently found useful is a TI-83 graphing calculator. Among other things, it can evaluate the frequency linearity of a variable tuned circuit. It is a miniature computer that can be used to run "basic" style programs away from home. A graphing calculator accomplishes the same thing as graphing software for a computer. The TI-83 has a computer interface so that graphs can be transferred to a regular computer. The cost of the calculator is generally less than the cost of software. But a special deal might be found on graphing software.

Calibration Standards

Many inexpensive instruments are not capable of being calibrated. DC and AC voltage standards for calibrating voltage instruments are particularly hard to come by. There are shunt regulator voltage standards available from semiconductor manufacturers. But these chips contain adjustment pins through which the device is supposed to be calibrated. Where does one find a standard to calibrate this semiconductor standard? I manage without having highly accurate voltage measuring instruments.

One area where inexpensive instruments are very accurate is in the realm of digital frequency counters. Since these instruments function by actually counting the peaks of the waveform measured, they are dependant only upon a time standard for accuracy. Accurate time standards can be produced by crystal oscillators. This allows for the frequency of an audio or RF generator to be very accurately set.

Resistors are commonly available in one percent tolerances. These can be used to calibrate ohm meters. Capacitors and inductors are not usually available with tolerances better than five percent. However, since frequency can be accurately measured, if either the inductance or capacitance of an L-C circuit is known, then the other value can be accurately determined by using a frequency counter and a mathematical calculation.

11 Complete Receivers

The Ocean Hopper

The Ocean Hopper is a three tube regenerative radio electronics kit that was sold by Allied Radio Corporation of Chicago back in the 1950s. There were earlier versions of the Ocean Hopper in the 1940s. The Allied Radio Corporation of today is located in Texas and sells electronic parts, but does not cater to electronic hobbyists. The Ocean Hopper discussed here is a three tube AC-DC design, one tube being a rectifier. The schematic is shown in Figure 11-1. I was a teenager in 1955 when I constructed an Ocean Hopper radio from the kit. I still have that radio and it is pictured in Figure 11-2.

The remarkable thing about this radio was that it offered general coverage from 150 kHz to 30 mHz using six plug in coils, reasonable RF sensitivity, and loud speaker output for a price of $15.50. This was the least

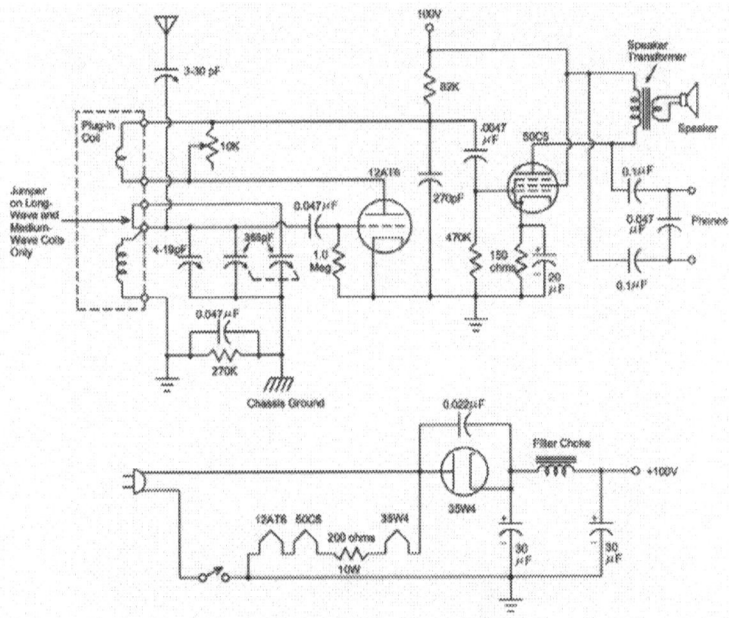

Figure 11-1: The Ocean Hopper Radio Schematic

expensive useful general coverage receiver available at the time. The receiver was to be used with an external speaker that was not supplied in the kit. External speakers were required with many shortwave receivers of that era.

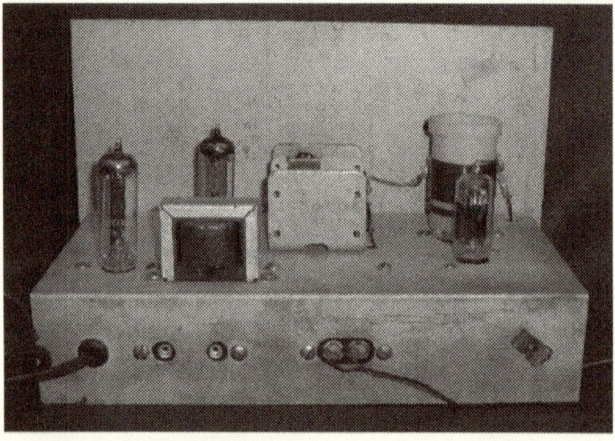

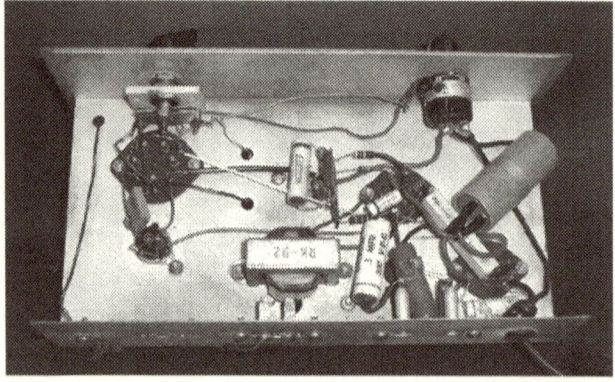

Figure 11-2: Three Views of the Ocean Hopper Radio

The Ocean Hopper had a superhet type bandset capacitor that covered a wide range of frequencies with each plug in coil. A small variable capacitor called a bandspread capacitor was connected in parallel with the bandset capacitor and was used to tune across a single shortwave broadcast band. This technique was used in regenerative ham band receivers of the day and it is similar to the band spreading technique used in this book where a fixed capacitor is used in parallel with a small variable capacitor and the received band is changed by switching inductors.

The antenna load on the tuned circuit of the Ocean Hopper would prevent the receiver from reaching the necessary oscillatory condition on some frequencies. To reduce the antenna load, the antenna was coupled to the tuned circuit through a variable 3 to 30 pF antenna coupling capacitor.

As good as the Ocean Hopper was, it suffered from many drawbacks. The radio did not have a calibrated frequency dial and determining what frequency was being received was a matter of experience with the shortwave bands and comparison with calibrated receivers. The regeneration control consisted of a shunt variable resistor across the tickler coil. The operation of this control was very touchy and noisy. The radio had no volume control and signals often came in too loud.

The Ocean Hopper was constructed on a metal chassis that was open on the sides. Fingers could easily stray in to the open sides and contact high voltage parts of the circuit. The power line was connected to the chassis through a parallel resistor-capacitor combination. This was supposed to make the chassis safe to touch, but it did not. The antenna was effectively isolated by the antenna tuning capacitor and the wires to the external speaker were isolated by the output transformer. But because of the lack of effective isolation of the chassis from the power line, electrical shocks received from this radio were frequent.

If this radio was constructed today, the filter choke, the speaker transformer, the plug in coil forms, and the variable capacitors would not be easy to find. The three miniature tubes are available at reasonable prices. The tube radio design in a following section is an attempt to duplicate the Ocean Hopper's good features with currently available parts, while eliminating the radio's shock hazard, and making improvements on many of its features.

The Space Spanner

The Space Spanner was a regenerative radio kit sold by Allied Radio that contained three tubes. One of the tubes, the 12AT7, was a dual triode. The radio was more convenient to use than the Ocean Hopper due to a built in speaker, a volume control, and internal band switching. However, the Space Spanner was not a general coverage receiver. The shortwave band covered 6.5 to 17 mHz. The radio could have been modified to provide wider frequency coverage.

The schematic diagram of the Space Spanner is shown in Figure 11-3 and a photograph appears in Figure 11-4. The two internal band determining coils were selected by a front panel switch. One coil was for the medium wave broadcast band and the other was for shortwave. The Space Spanner had a volume control, an extra stage of audio for more gain, and an internal speaker. It sold for $16.00 in 1958.

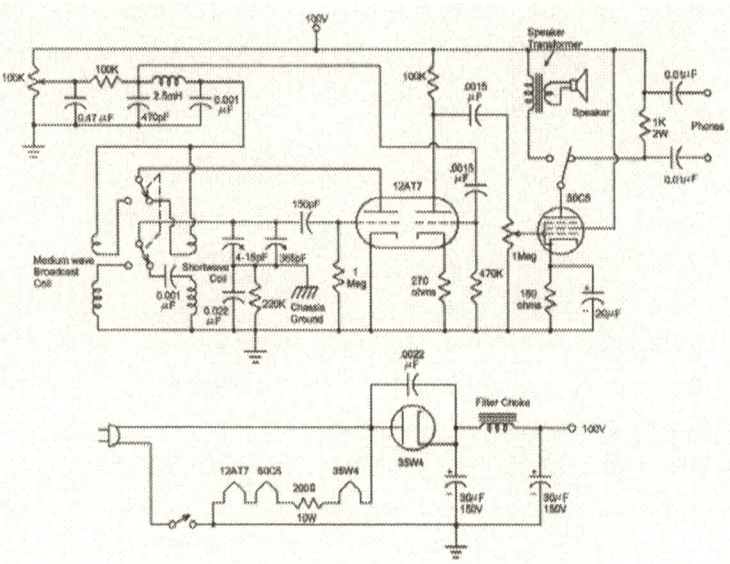

Figure 11-3: The Space Spanner Radio Schematic

The Space Spanner was housed in a wooden cabinet and it was somewhat safer than the Ocean Hopper. But touching the metal front panel or the headphone terminals could result in a shock.

Figure 11-4: The Space Spanner Radio

The Globe-Span Modern Transistor/IC Regenerative Radio

The transistor regenerative radio presented here was featured in the June, 2002 issue of *Nuts and Volts Magazine* where it was called the Globe-Span Radio. The design has been modified since the publication because two of the parts originally specified became difficult to obtain. The radio schematic is shown in Figure 11-5 and information on parts is given in the accompanying Parts List. Two photographic views of the radio are shown in Figures 11-6A and 11-6B. The design uses bipolar and FET transistors, one analog IC (the audio power amplifier), and two power regulator ICs. As mentioned previously, a regenerative radio can be built with a single transistor. The added complexity of this radio allows for improved performance, more versatility, and speaker output.

The high impedance antenna input can be used with a short wire antenna which will act like an active antenna. A medium wave high pass filter (C9, C10, R4, and R5) is placed between the high impedance buffer and the input of the FET regenerative amplifier. This filter will not be necessary when the radio is operated away from strong medium wave stations. The low impedance antenna input does not have the medium wave filter. If only the low impedance input is desired, the transistor Q1 and the associated circuitry can be eliminated.

Note that the antenna is completely isolated from the tuned circuit (L1-L6, C18, and C19). Thus, the characteristics of the antenna have little effect on the frequency of the tuned circuit. However, a strong signal can effect the tuning

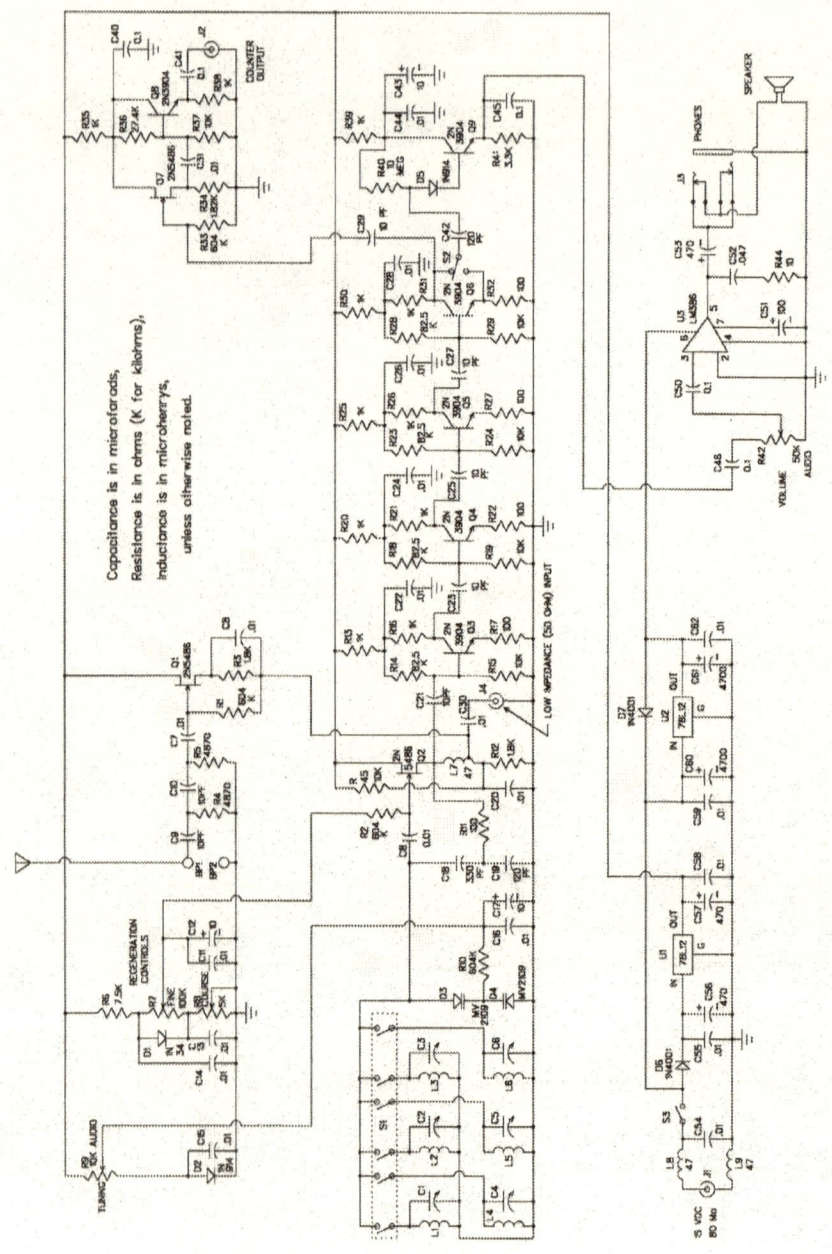

Figure 11-5: Schematic Diagram of the Globe-Span Receiver

PARTS LIST

Semiconductors

D1	1N34
D2, D5	1N914
D3, D4	MV2109
D6, D7	1N4001
Q1, Q2, Q7	2N5486
Q3, Q4, Q5 Q6, Q8, Q9	2N3904
U1, U2	78L12
U3	LM386

Resistors

R1, R2, R10, R33	604K metal film
R3, R12	1.8K
R4, R5	4870 metal film
R6	7.5K
R7	100K linear pot
R8	5K linear pot
R9	10K audio pot
R13, R20, R25 R30, R35, R39	1K
R14, R18, R23, R28	82.5K metal film
R15, R19, R24 R29, R37, R45	10K metal film
R16, R21, R26 R31, R38	1K metal film
R11, R17, R22 R27, R32	100 ohm metal film
R34	1.82K metal film
R36	27.4K metal film
R40	10Meg
R41	3.3K
R42	50K audio pot with switch
R43	Not used, disregard space on printed circuit board.
R44	10 ohm

PARTS LIST (continued)

Capacitors

C1 - C6	3.5 to 20pF ceramic trimmer
C7, C8, C11, C13, C14, C15, C16, C20, C22, C24, C26, C28, C30, C31, C44, C54, C55, C58, C59, C62	0.01 µF ceramic disk
C9, C10, C21, C23, C25, C27, C29	10 pF silver mica
C12, C17, C43	10 µF 16 volt electrolytic
C18	330 pF silver mica
C19, C42	120 pF silver mica
C32 to C39	These numbers not used
C40, C41, C45, C46, C50	0.1 µF ceramic disk
C47, C48, C49	These numbers not used
C51	100 µF 16 volt electrolytic
C52	.047 µF ceramic disk
C53, C56, C57	470 µF, 16 volt electrolytic
C60, C61	4700 µF, 16 volt electrolytic

Inductors

L1 to L6	Select for the desired bands, See Tables 1-1 and 1-2.
L7	29 turns of #24 wire on an FT-50A-61 toroidal core. Tap 4 turns from one end.
L8, L9	29 turns of #24 wire on an FT-50A-43 or FT-50A-61 core

Miscellaneous

J1	DC power jack
J2	Jack to connect frequency counter
J3	Headphone jack
J4	Coaxial jack for Low impedance (50 to 75 ohm) antenna connection
JU	Solid wire jumpers on PC board
BP1, BP2	Antenna binding posts
S1	6 pole dip switch
S2	SPDT miniature slide switch
S3	Power switch on R42
SPKR	4 inch 8 ohm speaker
DC power adapter	15 volts 500 mA wall power

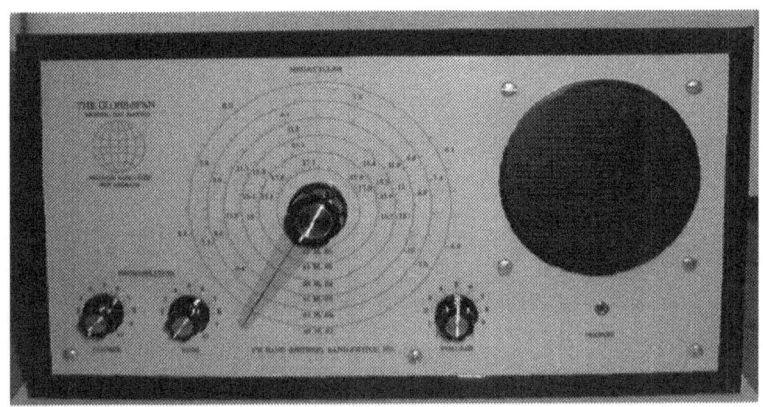

Figure 11-6A: The Globe-Span Radio Front View

Figure 11-6B: The Globe-Span Radio Rear View

characteristics and the point of oscillation of the radio. The load caused by an antenna connected to the low impedance antenna input will somewhat effect the amount of regeneration required to make the receiver oscillate.

The Colpitts design is used for the regenerative amplifier, Q2. A list of inductor values that can be used for L1 through L6 is given in Tables 1-1 and 1-2. Standard commercial inductance values are given in Table 1-2 and commercially manufactured inductors can be used if desired. But hand wound toroids will have a higher Q factor, provide a better performance, and cost less.

The three considerations in choosing inductors are the inductance value, the self resonant frequency of the coil, and the Q factor. The value of the inductance determines what band of frequencies the radio will receive using that coil. The self resonant frequency of the coil must be much higher than the highest frequency to be received by the radio when using that coil. A higher Q factor of the coil leads to narrower tuning but commercial coils with a Q of only 40 have been used successfully in this circuit. If the tickler coil or the Hartley design was used, the coil for each band would have to be individually designed with attention paid to the feedback ratio. Also, a multi-pole switch or plug-in coils would have to be used to change bands.

In the Colpitts circuit, the feedback ratio is determined by the capacitors C18 and C19. A JFET (Q2) is used for the regenerative amplifier. Extra wideband RF gain is desirable after the regenerative amplifier stage. This is provided by the four transistor stages Q3, Q4, Q5 and Q6. The fourth stage can be switched between gains of 1 and 10 (0 DB and 20 DB). Thus, the total gain of the wideband amplifier can be 60 DB or 80 DB. A gain of 60 DB is the setting that is most often used. The 80 DB setting can be useful on the higher frequency bands (15 mHz and above) during local daytime when signals are weak. Transistors Q7 and Q8 provide an output for a frequency counter. When a counter is connected to the receiver, the frequency of a strong signal can be directly determined.

The AM detector circuit (Q9) is designed to reduce the load on the preceding amplifier stage and to work with a signal source that is not at DC ground. An attribute of this detector is that the diode can be silicon instead of the usual point contact germanium type. That is because this detector compensates for the 0.7 volt offset of a silicon diode.

The power amplifier is a LM386 IC that provides 300 milliwatts output to an 8 ohm speaker. The LM386 is simple to use and requires no heat sink. There are versions of this IC that offer one watt output, but the 300 milliwatt version is adequate for this application. The desired speaker is a 4 inch (or larger) 8 ohm unit. The power amplifier provides good audio with a hint of bass output. As seen in the photographs of Figures 11-6A and 11-6B, the radio was constructed breadboard style. It was given an old fashioned appearance and the old sounding name, "Globe-Span".

The overall performance of the radio compares favorably with shortwave superheterodyne receivers. However, superhet shortwave receivers can filter out strong signals near the frequency of the desired station that the Globe-Span is unable to completely eliminate. Even so, the Globe-Span regenerative radio will bring in most international shortwave stations and it is a pleasure to work with.

The printed circuit board pattern for the Globe-Span is given in Figure 11-7 and the parts placement diagram is in Figure 11-8. Etched and drilled PC boards and complete parts kits are available from the author[3]. The front panel design was drawn with a computer drawing program and printed out on legal size paper. The front panel layout appears in Figure 11-9. The printed paper was laminated and then glued to the particle board. Poster edging was used to cover the edges.

The frequency dial markings for Figure 11-9 were made by measuring the tuning angle for different frequency locations. The frequency marks were then drawn on the dial by placing them at the required angle. Since varactor tuning was used with an audio type of potentiometer supplying the tuning voltage, the dial is nonlinear near the middle. As previously mentioned, this is because the characteristic of an audio potentiometer is not a true exponential curve.

Further Modifications and Improvements of the Transistor/IC Radio

A linear varactor diode voltage circuit that can be used to achieve a linear frequency versus control shaft rotation has been previously mentioned (Figure 9-3).

In the initial design of the Globe-Span, all controls except the band switch were DC voltage controls. The band switch was mounted directly on the PC board and all signals stayed on the circuit board from the antenna to the speaker wires. All functions of the radio except band switching could be remote

---

[3] The author has for sale a limited number of etched and drilled circuit boards and parts kits for the Globe-Span. The price is $17 for a PC board and $89 for a kit of all <u>electronic</u> parts. Shipping and handling for the continental U.S. and Canada is included. At the publish date of this book, the author's address was HCR 2 Box 2104A, Van Buren MO 63965 and the email address was lyle009@juno.com.

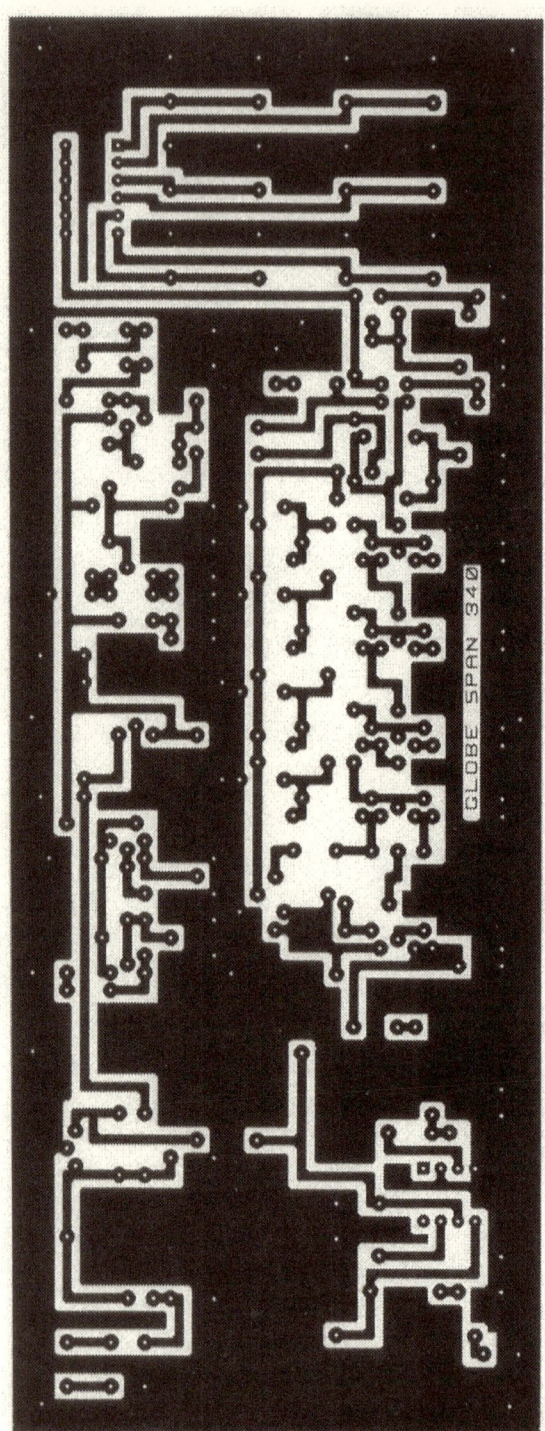

Figure 11-7: Printed Circuit Foil Side Pattern for the Globe-Span Radio

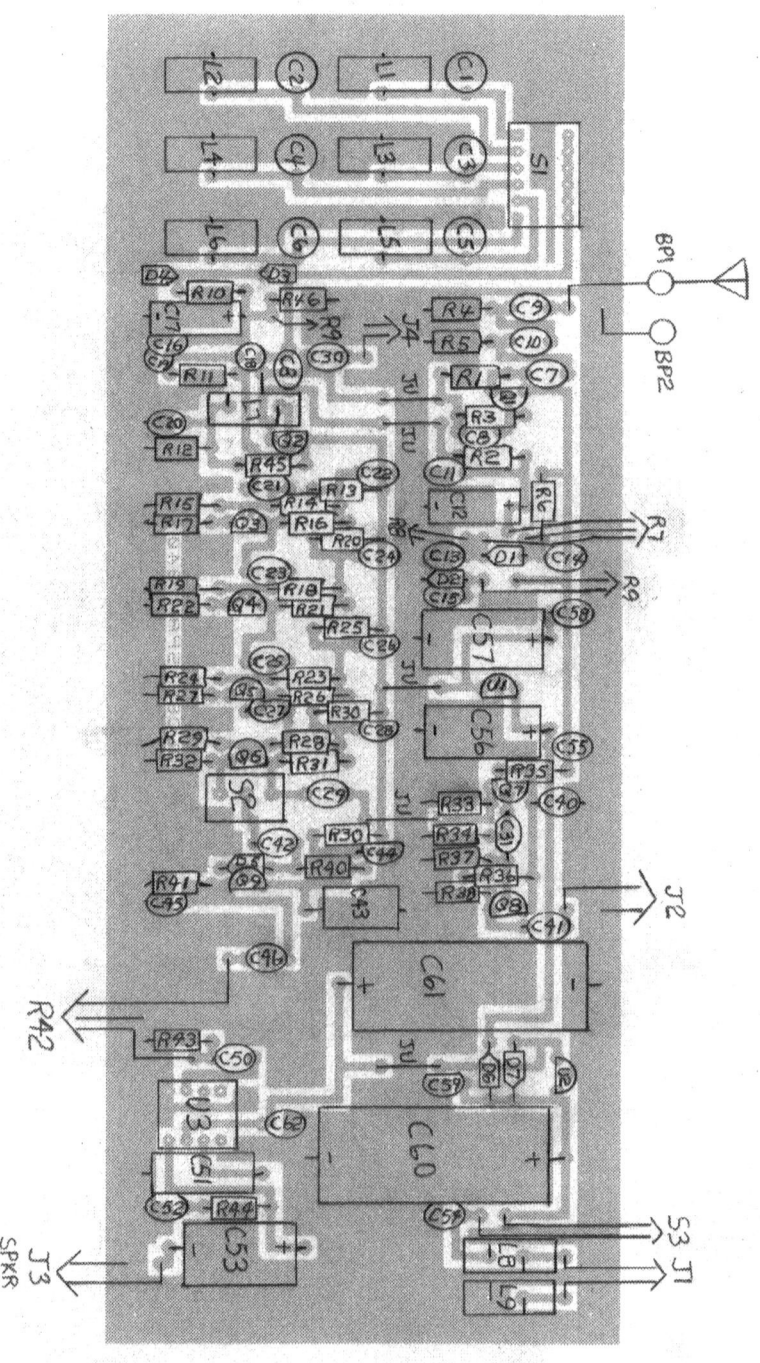

Figure 11-8: Parts Placement Diagram for the Printed Circuit in Figure 11-7

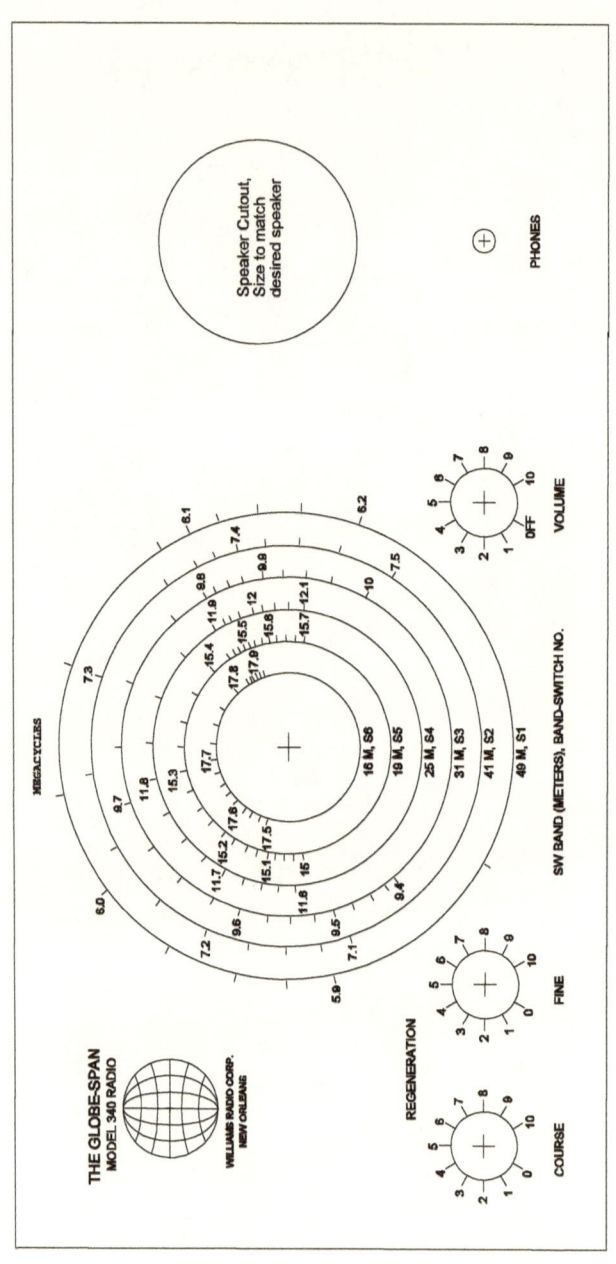

Fgure 11-9 Front Panel of the Globe-Span Radio

controlled through four DC control lines. The original receiver could be controlled by a personal computer or by microprocessors such as the PIC or Basic Stamp. Four digital to analog (D/A) converters would be required to interface a computer with the radio. Readers that like digital projects may be interested in this possibility.

The volume in the original design was controlled by the MC3340P electronic volume control IC, which is now unobtainable. In the present design, a conventional off board volume control is used.

Newly Designed Vacuum Tube Regenerative Radio

As previously mentioned, the only reason for building a tube radio is the exotic quality of the receiver and the historical perspective provided by such a project.

While writing this book, I have been working on new tube regenerative radio designs. Since the original regenerative radios were the tube type, it would seem that design of a new one would be straightforward. However, this is not the case. Tube regenerative radios are more likely than transistor radios to have instability problems.

It is desired that the oscillating regenerative radio produces only one unmodulated radio frequency. Tube circuits tend to produce a radio frequency that is modulated at one or several audio frequencies. As discussed elsewhere, this phenomena is called squegging. I have not yet been able to stabilize the tube version of the Colpitts circuit that can be used with two terminal inductors. The Hartley circuit shown in Figure 11-10 has been stabilized.

The regenerative tube in this circuit may need to be shielded. A tube shield is a cylindrical metal sleeve that covers the side of a glass tube to keep out electrostatic and magnetic fields. Because of this requirement, it may be as well to select an octal metal tube that comes with a built-in shield.

The tube regenerative circuit shown in Figure 11-10 is not necessarily the best three tube design possible, but it has several improvements over the Ocean Hopper and Space Spanner. All of the improvements made in the Globe-Span radio have been incorporated into this tube version. I expect that further improvements will be possible.

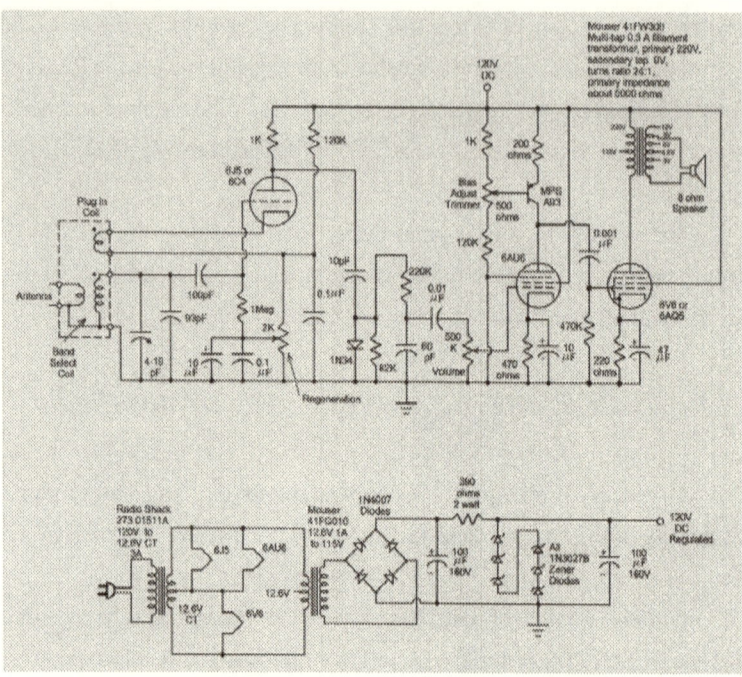

Figure 11-10: Schematic of the Tube Regenerative Radio

One feature of the radio is that it is double isolated from the power line by the back to back power transformers and the shock hazard has been eliminated. The Ocean Hopper has a gain of around 25. The gain of the 6AU6 tube in Figure 11-10 is around 250, an increase of ten times or 20 decibels. The radio here uses grid bias regeneration control, which is smooth and free of hysteresis. Figure 11-11 is a photograph of a radio that is nearly the same as the radio in the circuit of Figure 11-10. The center tube in this photo is not a 6AU6.

As with the transistor design, there is a small variable tuning capacitor across a large fixed capacitance so that the total tuning range is narrow with respect to the center frequency of the band. At the lower shortwave frequencies, the width of the tuning range is about the width of a shortwave commercial band (about 300 kHz). On the higher frequency shortwave bands, the tuning range will be about twice the width of a commercial band.

Because this receiver requires coils that have multiple connections, plug in coils are used instead of a band switch. Table 1-2 can be used to build the main tuning inductor, but the feedback and the antenna windings will have to be determined by trial and error. For the 19 meter band, the main inductor has 13

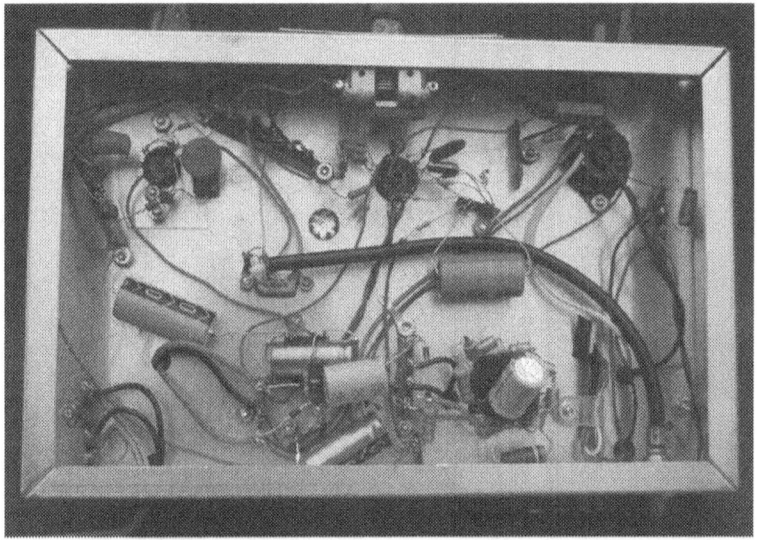

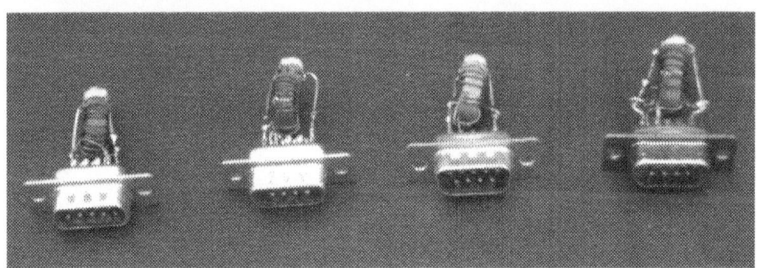

Figure 11-11: The Vacuum Tube Regenerative Radio With Plug-in Coils

turns, as shown in Table 1-2. The feedback winding is 4 turns and the antenna winding is 1 turn. This receiver does not have a high impedance antenna input in its present configuration.

Since the first tube has been biased to operate in the linear range, detection does not occur in that tube. Detection has been accomplished by a detector network containing a germanium diode which is placed between the regenerative stage and the audio amplifier tube.

If a 4 to 20 pF variable capacitor is not available for this project, the varactor circuit (potentiometer R9, varactor diodes D3 and D4, and resistor R10) of Figure 11-5 can be used. An extra 12 volt DC power supply will be required for varactors. The 12 volt supply might be derived from the filament supply AC voltage. The current drawn by the varactors is very small, less than a microampere.

As discussed in Chapter 5, an interim audio amplifier is used between the first tube and the power output tube. In spite of the theoretical advantage of using an interim RF amplifier, the use of an audio amplifier here works very well. The transistor in the plate circuit of the 6AU6 tube is not an amplifier, but rather an active load that supplies the 3.7 ma DC plate current while presenting a high load impedance to the tube. This high plate circuit impedance allows the tube to have a much higher gain than would otherwise be possible. The audio power amplifier shown will supply about 450 mW audio output.

This receiver contains three tubes plus a silicon rectifier, so the radio is equivalent to a four tube AC radio or a three tube battery radio as they would have been built in the 1950s. Therefore, this design effectively contains one more tube than the Ocean Hopper and Space Spanner.

The overall gain of this tube receiver is about ten times the gain of the Ocean Hopper, but less than the transistor version in Figure 11-5.

www.ingramcontent.com/pod-product-compliance
Lightning Source LLC
Chambersburg PA
CBHW031920240526
45464CB00021B/594